No Need to Count

No Need to Count

A Practical Approach to Casino Blackjack

Leon B. Dubey, Jr.

San Diego • New York
A. S. Barnes & Company, Inc.
In London:
The Tantivy Press

The Tantivy Press
Magdalen House
136-148 Tooley Street
London, SE1 2TT, England

First Edition
Manufactured in the United States of America
For information write to A. S. Barnes and Company, Inc.,
P.O. Box 3051, San Diego, CA 92038

Library of Congress Cataloging in Publication Data

Dubey, Leon B
    No need to count.

    Bibliography: p.

    1.   Blackjack (Game)   I.   Title.
GV1295.B55D82   795.4'2   79-23884
ISBN 0-498-02465-2

1 2 3 4 5 6 7 8 9   84 83 82 81 80

To Mr. and Mrs. Leon B. Dubey, Sr.

CONTENTS

PREAMBLE

As I write this, a controversy of major import to the casino blackjack enthusiast is raging wildly. In fact, the onslaught of knowledgeable card counting system players has alarmed casino management to such an extent that they have actually banned this sort of player in Las Vegas and certain other places. Considerable dialogue is now appearing in the national media pertaining to a proposal by New Jersey casino operators to further this dastardly practice by ejecting recognizable card counters.

This book contains the optimum solution to this problem! You see, with *No Need to Count*, you have a winning system that is not recognizable as card counting because in fact, you do not have to count cards to win!

As you exercise my system in the casinos, you are marching to a different drummer. You appear to be almost haphazardly betting without any awareness of anybody's point count method. My wife, myself, and several relatives and friends got away with playing my way in the casinos for many thousands of hands before our collective accrued winnings attracted the management's interest. At the same time, the casino people were already painfully aware of their vulnerability to counters. If we can get away with it, so can you!

Want some very important advice? Keep your variation in bet size at less than three or four to one. Widely varying bet sizes that are strangely correlated to winning are the loudest alarm you can set off. A Las Vegas casino manager once told me that he would eject anyone who wins consistently with a maximum to minimum bet ratio in excess of four to one. The way things are now, I would not exceed two to one very often.

Let me offer you one more equally significant tip. Never stay in any one place, or at any one table, for more than a half hour or so. Do not give them a chance to catch on to your smarts. Try to work in teams of several players and mix up your sequences of who plays where and when, constantly. Take their money and run. If you are too greedy, the curtain will surely fall on you.

About the only way they can stop you with this technique would be flatly to disallow anybody who wins at all to continue to play. If they do that, who would play? If they change the rules significantly enough to defeat you, they will defeat themselves by turning off everyone's continued interest in the game and, therefore, their own livelihood. They have a problem. Get in on the action before they find a solution.

My system is ridiculously simple and it works! Let them ban the counters. With *No Need to Count*, you can, indeed, overcome!

ACKNOWLEDGMENTS

A number of people have assisted me during various portions of the six-year development period for this book. My wife, Phyllis, helped out with every phase. In particular, she manually played and recorded twenty thousand hands of blackjack with me, laboriously calculated a mountain of statistics, participated heavily in the casino tests of my "Complete System," actually conceived of the Du-Rite Wheel, carefully wrote out and checked endless computer input forms, typed and retyped the manuscript and served as a much-needed sounding board for every concept.

I have also received considerable help from Dave Anderson, trajectory analyst; Dr. Mikiso Mizuki, statistician; Stuart Keith, programmer-analyst, Bruce Carlson, ship's programmer (Apollo ship U.S.N.S. *Vanguard*); Al Robertson, ship's programmer (Apollo ship U.S.N.S. *Huntsville*); Harold Feldstein, computer scientist and bridge authority; Dan Shaughnessy, senior programmer; and Paul

Fiorito, Central Data Processing group leader (Apollo ship U.S.N.S. *Vanguard*).

I would also like to express my appreciation to Susan Armstrong for considerable calculations, and to George and Mary Sykes for developing my interest in the game and participating in the casino testing of my theories.

Finally, I am indebted to the manager of the former Bonanza Casino, whose wish for anonymity shall be respected, for our very illuminating and honest discussions on systems players, casino countermeasures and cheating.

No Need to Count

1. INTRODUCTION

Few men have analyzed the results of as many hands of basic casino blackjack as I have. I do not make this statement as a braggart. Rather, I wish to establish that what you will read here is based on scientific facts garnered from years of applied research. You will find the following supporting evidence in this book:

1. The results of many millions[1] of blackjack hands generated by my own computer program.
2. A summary of twenty thousand hands played by hand under simulated casino conditions.
3. Documented results of over ten thousand hands of actual casino testing of my unique and winning method. This system is published here for the first time.

As the title and subtitle suggest, this book is devoted to the practical blackjack player who wishes to employ a simple, workable system that gives him or her a significant edge on the house. To my mind, this means a system that does not require you to keep track of what cards have been played or to memorize a variable strategy.

Does this sound impossible to you? Well, let me and my multi-million blackjack hands assure you that such a method is not only possible, but available to you in this very book. See chapter 5 for a very pleasant surprise.

To apply this technique you need only learn the fixed playing responses (basic strategy) of chapter 3 and a few simple betting rules. The full-blown strategy can be learned in an hour or two with the aid of the Du-Rite Wheel described in this book. An *adequate*, but less precise, *reduced strategy* is also presented in chapter 3. You can memorize this in about ten minutes, I should think. Furthermore, you can take your Du-Rite Wheel with you to the casino as a handy, and generally permissible reference. The betting rules are learned by reading them once; they are that simple.

Of course, the advent of the electronic computer has caused a widespread revival of the analysis of many casino gambling games. Blackjack, or twenty-one, has suffered a particularly heavy onslaught. Such efforts have been largely devoted to two aspects of the game. These are (1) the development of a "basic strategy" that, when followed, will produce a slight edge in favor of the player, and (2) the development of various "card counting" methods to be employed for optimum bet sizing and that give the player a significant edge.

You will find the correct strategy in chapter 3. Here also are the summarized results of the computer implementation of this strategy by myself and many other computer scientists. A discussion of the various card counting methods is undertaken in chapter 9, should you be interested in these more sophisticated and difficult techniques.

Unfortunately, the "card counting" systems are not so easily implemented in the casino. To employ these methods effectively, a person must have the razor-sharp, lightning-fast mind of the near genius, and he must subject himself to many hours of rigorous and dry training.

Anyone who has ever played blackjack on the "Strip" in Las Vegas will attest to the speed at which the game is played. The dealers are so fast that their hands are almost a blur. It is extremely difficult under these conditions to assign a point value to each card as it falls, keep a running count of the cards remaining in the deck, divide one by the other to obtain a numerical ratio, refer to mentally stored tables based on this ratio and vary the playing strategy accordingly, and at the same time determine and place the optimum bet size. An unusually agile mind, combined with long hours of training, is a must for such methods.

This book has been written for the millions of players who cannot, or will not, meet the above qualifications. Optimum systematic betting techniques are presented in chapter 5 that have been proven to *win* predictably. Yet they can be learned by *anyone* in just a few seconds. Significantly, these methods do not involve memorizing exposed cards in any way, nor do they require a variable strategy.

Now, a lot of people have proposed all sorts of gambling systems that can be applied to blackjack and that will not require you to count cards. The most famous of these is the "double your bet when you lose" system, or the Small Martingale, as it is also known. This system, along with all the known and little-known systems that my research turned up, has been applied to a hand-generated sample of twenty-thousand hands for comparison purposes.

In chapter 8, then, you will find graphical twenty-thousand-hand histories that debunk all progression systems such as the double-up, Grand Martingale and the Labby. Many other systems are investigated as well, and profit versus the number of hands played is graphed for them all. Surprisingly enough, some of these may have a little merit, from a money management standpoint, when applied to blackjack, roulette and craps. See chapter 8 for this type of information. It is a good bet that you will find your favorite system treated there.

Also featured are a number of powerful "tools" for the blackjack player who wants to do some home testing of a system. A set of cumulative *profit coefficients* are presented that represent twenty thousand hands of blackjack in five-hundred-hand steps. These coefficients may be used to obtain a historical profit record of most types of betting systems in just a few minutes of calculations. Thus, a twenty-thousand-hand comparison of a system developed by the reader may be obtained very quickly employing the methods of chapter 8. To use these coefficients, you need only know how to multiply.

Another gambler's tool developed in the book is a complete breakdown of how to gather and record your own blackjack data. Exactly what statistics should be recorded and how they should be collected are discussed along with examples. A set of elementary formulas for calculating the profit are presented in chapter 7 and derived in Appendixes B and C.

Chapter 6 will enable you to effectively size your bet to your bankroll and will terrify you with tales of observed statistical fluctuations in capital. To your terror, you may add horror and considerable caution by reading chapter 10, which is devoted to casino countermeasures.

Since I wrote this book for everyone, not just for mathematicians and scientists, I have taken great pains throughout to keep all discussions and examples simple to understand and pleasurable to read. It is sincerely hoped that everyone who reads this book will find the process enjoyable and profitable.

NOTE

1. The quoted percentages of chapter 4 are based on a five-million-hand computer run. However, I have generated tens of millions of hands with this program and obtained essentially identical results.

2. HOW BLACKJACK IS PLAYED

GENERAL DESCRIPTION

Blackjack, or twenty-one, is a popular gambling card game widely enjoyed in the home and casino. It is the *casino* game with which this book is concerned, although variations of many of the winning techniques to be discussed could be exploited in the *home* game as well.

Legal casinos offering blackjack exist in profusion in the state of Nevada, particularly in the Las Vegas, Reno and Lake Tahoe areas. A new gambling center is now in its infancy at Atlantic City, N.J. Isolated games, mostly illegal, are located at various spots in the continental United States; but Nevada is the undisputed mecca for American blackjack. San Juan, Puerto Rico, has about a dozen casinos offering blackjack; and legal casinos, usually no more than one to an island, may be found in the Grand Bahamas, West Indies and the Philippines. The small island of Aruba in the Netherland's Antilles has a half dozen casinos with a Las Vegas atmosphere.

In Europe, the game is primarily known as vingt-et-un (French for twenty-one), or "Van-John," a popular British corruption of this. It is played in legal casinos and clubs in London, Venice and on the Isle of Man. "Pontoon" is another popular alias for blackjack—probably derived from the fact that the English and Australians often refer to the players as "punters." A description of the game that employs the term "punters" frequently may be found in Hervey's book[1] should you be curious about such things.

The game is played at plush blackjack tables covered with the inevitable green felt. The layout of the table and arrangement of the dealers and players are sketched below.

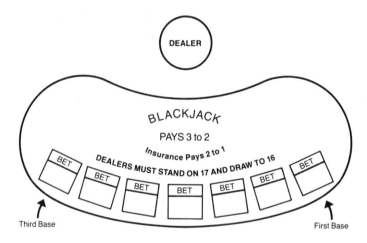

Typical Blackjack Layout

The players sit as shown, equally spaced around the convex side of the curved table. The dealer stands centrally on the concave side and dispenses the cards to the players from his left to his right. The seat at the extreme left of the dealer is often referred to as "first base" while that at his extreme right is known as "third base." For many card counting systems, it is best to sit as close to third base as possible so as to maximize the number of cards seen before you play. For my systems, however, your position makes little or no difference since you are not interested in anyone else's cards.

In most establishments, the gambling tables are arranged in one or more ovals or rectangles. Each loop of tables, which may include craps and roulette setups as well as blackjack tables, encloses the "pit" area. This pit area is off limits to all players and harbors such employees as "pit bosses," dealers, croupiers, stickmen and drink girls.

As you might expect, the pit boss oversees the other employees and is in complete charge of his area. He watches the progress of both players and dealers and may direct some sort of action, or countermeasure, if someone appears to be winning too consistently. Some pit bosses are reputed to remember the faces of hundreds of gamblers who are dangerous to the house. My face, and my wife's, have unfortunately become two of these upon occasion in Las Vegas and other casinos.

Normally pit bosses and dealers are reasonably tolerant individuals, but, as in the case of my friend Pat Munn, they will occasionally clamp down on a troublesome customer. Actually, Pat is far from a troublesome person, but she did have a very unfortunate experience. I will let her tell you what happened.

I was playing blackjack in Caesar's Palace, at a crowded table, when I inadvertently upended my drink with my furpiece. Of course, I was highly embarrassed, but the dealer was very nice about it. He said something like, "Oh, forget it, it's nothing," and had the pit boss signal for a drink girl while he sopped up Scotch from the green felt. When my new, free Scotch arrived, I thanked everyone and sheepishly got up and left the table.

Not wishing to attract any further attention I wandered, drink in hand, to the nearby roulette wheel where a group of Chinese, maybe six or eight of them, were animatedly executing a wild system. I had no sooner arrived in the midst of this minor pandemonium, when I was slapped heartily on the back by my friend Jerry, as he gleefully shouted, "Guess how much I just won?" To my utter mortification, the impact of the blackslap caused the full drink glass to shoot out of my hand and land with a sickening crash in the middle of the spinning roulette wheel! A fine spray of Scotch rained down upon the sputtering Chinamen and I was struck dumb! I stood there in shock, with my drink hand still upraised, as the same pit boss walked over and asked me firmly if I'd like to leave. In an embarrassed show of bravado, I stupidly cracked, "Okay, I've been thrown out of better places than this, you know," and numbly started for the door. The pit boss of the fabulous multimillion-dollar casino calmly answered, "I doubt it" as I hastily retreated.

Pat found out later that they had to dismantle the wheel and clean it up before it could be used again.

The "drink girls" are generally signaled by the pit boss and offer unlimited free drinks to all players. This practice is universal in Nevada but just about nonexistent everywhere else. You can get a free drink or snack in Aruba by asking for them.

Croupiers and stickmen are not involved in blackjack.

During actual casino play, from one to seven players compete individually against a dealer. The object of each player is to obtain cards whose total is as close to 21 as possible without exceeding it. If the player's total is less than the dealer's, or over 21, the player loses. If the player and the dealer have identical totals of 21 or less, the hand is considered a standoff or "push." No money changes hands on a push.

Before the cards are dealt for each hand, each player places his bet in the little rectangle on the felt directly in front of him. After all the players and the dealer have received all their cards, the dealer turns over everyone's cards and proceeds to either pay or collect from each player. If you have won, he will place an amount equivalent to your bet in your box. If you have been beaten by him, he collects your bet and places it in the chip rack. Should you have a push, the dealer will acknowledge this by sharply rapping the table in front of you.

Incidentally, almost all betting is done with chips. Actual cash is rarely used in the casino. As you sit down at a table, you purchase as many chips as you desire. If these become exhausted, you "dig" for money to purchase some more. The most commonly used chips are valued at one dollar and five dollars. Chips of a higher value are rarely seen at the blackjack tables. Chips may also be purchased or redeemed for money at the cashiers' windows in each casino. Metal one-dollar chips are not accepted by any but the issuing casino. Five-dollar chips, however, can be used anywhere in Las Vegas.

Most casinos in Nevada have tables that employ a single deck of fifty-two cards. In downtown Las Vegas and Reno, for example, the single-deck game prevails, with most houses also offering one or more tables that use two decks shuffled together. On the Las Vegas "Strip," however, the multideck game heavily dominates the scene. Two, and more often four or more, decks shuffled together and dealt from a container called a "shoe" represent a large percentage of the tables on the "Strip." Most often, however, each house has at least one table open offering a fifty-two card game. Normally these single-deck tables will have a five-dollar minimum bet sign displayed and, as a consequence, are usually fairly empty. For the single-deck games in downtown Las Vegas, the betting minimum is sometimes fifty cents during the week and one dollar on the weekend. The maximum bet values are predominantly $200 or $500 in Las Vegas.

Single-deck games also prevail in Reno and Lake Tahoe, but are otherwise quite difficult to find. In Puerto Rico for example, two decks are the mainstay, while European casinos lean almost exclusively toward the multideck game, with four or more decks being the most common. Six deck shoes predominate in Aruba.

The reason that the multideck game is so prevalent is that the player's overall basic advantage is moderately reduced under these circumstances. Moreover, the effects of most systems that actually win (these will be described at length in succeeding chapters) are also somewhat damped out when more than one deck is used. Essentially, though, multidecks will simply result in a slower win rate, since such a game ultimately reduces to a one-deck situation, on the average, after sufficient cards have been played. Betting approaches to accommodate the multideck situation are discussed in chapter 5.

A huge variable in any casino is the attitude of the dealers. Most dealers are really very friendly. In fact, if you appear to be a complete novice, you will be treated in a typically fatherly fashion. Dealers exude friendliness and are extremely helpful when you are a fresh new customer. Sometimes they are so helpful it is hard to believe.

In the fall of 1964, my wife and I made our first trip to Las Vegas. Neither of us had ever played blackjack before and we knew absolutely nothing about the game. When Phyllis sat down to her first table she was one big mistake! She did whatever anyone told her (most often the advice came from the dealer) and could not even add her hand totals fast enough to know what was happening. Yet she could not seem to lose! She won, and won, and won, sometimes as many as eight or nine hands in a row. But she was only betting fifty cents or a dollar a clip and actually accumulated very little winnings.

Meanwhile, the dealer was beside himself with helpful hints and personal joy. Phyllis could not seem to lose her fifty-cent bets despite her naiveté, but an academic-looking big bettor, some two seats away at the same table, was losing a fortune. Oh how the dealer loved it! "Look at that," he chirped gleefully. "Another blackjack for the little lady. How about that! Oh my goodness sir. Another ten for you. What a shame. What a shame."

Of course, this affair reeks of hanky-panky. But it was conducted so pleasantly! It is not at all unusual for a winning player to be treated quite nastily by an unhappy dealer. See chapter 10 for some blatant instances of this.

This completes a general picture of the typical casino environment and playing setup for blackjack. It remains to specify the simple ground rules and available player options in order to complete our understanding of the mechanics of the game. This is next accomplished.

TYPICAL CASINO RULES

Shuffling, Cutting, Burning, and the Pickup

When the table first opens, or the deck is exhausted, or a new deck is brought in, the cards are thoroughly shuffled by the dealer. Naturally, if the deck runs out in the middle of a hand it is shuffled, exclusive of those cards currently on the table. It is, at the time of this writing, also legal for the dealer to shuffle at any time he or she pleases, whether the cards have run out or not. Moreover, the dealer will not hesitate to engage in this practice, popularly termed "shuffleup," if he feels it will help him to defeat a winning system player.

Nevertheless, most houses instruct their dealers to go to the end of the deck or, much more frequently, to some specified remaining number of cards, say fifteen or twenty, before shuffling. Another typical house rule, especially in Reno, is to shuffle when it is evident that all players cannot otherwise complete the next round of hands. In actual play, personal experience has shown that the dealer will increase his shuffling frequency somewhat when he discovers a winning system player, but rarely enough to slow the game down to an utterly boring pace. Since, as will be discussed in succeeding chapters, it is advantageous to the house to discretely increase its shuffling frequency, particularly against card counters, the casino is faced with a trade-off between losing customers through boredom and losing money to the system players. With every passing year, the dealers get less and less reluctant to shuffle at the slightest provocation, despite the fact that the extra time consumption can be costly.

Although I have never seen it done in all my casino experience, you can ask for a shuffle between hands. The dealer may, or may not, accommodate you, depending upon his opinion as to why you wish it, and also upon whether or not he is cheating. If he thinks such a shuffle may help your system, which is indeed a possibility, or if he has nicely stacked the deck against you, always a distinct possibility, your request may not meet with much enthusiasm. If, on the other hand, he thinks you are a typical losing jerk, he may shuffle at your command with little hesitation.

Upon completing a shuffle, the dealer will offer the deck to a player at random, whoever strikes his fancy, so that it may be cut. In many casinos an extra card, not a playing card, will be placed alongside the offered deck. The chosen player will cut a portion of the deck off the top (five or more cards are technically required) and place this portion on top of the extra card. The dealer will then complete the cut by placing the remaining cards on top.

Alternatively, the dealer may extend a full deck out to a player who will effect the cut by sliding the extra card somewhere into the deck. A third possibility is just a standard cut with no extra card.

Which custom is employed depends upon the casino. I only mention these variations for completeness, and to remove as many elements of the unknown as possible from the minds of the readers and future players.

After the cut, a card is "burned." That is, a card, usually the top one, is placed faceup on the bottom of the deck. In actual practice, no particular effort is ever made by the dealer to show this card to the players. In fact, many dealers deliberately try to obscure this card from view as a further attempt to hinder card counters. Of course, employing my systems, you will not be very concerned about this since they do not require any a priori knowledge of exposed cards.

After a complete hand, the dealer will pick up all the cards and either place them on the bottom of the deck, or in a little rectangular receptacle. The receptacle is now used more frequently than the bottom of the deck. I suppose this is meant to be in deference to the player since it offers less opportunity for the dealer to perform slight of hand. Actually, the container does not help much in this regard. If the dealer wants to cheat, a little thing like that will never stop him. He can stack the cards against you *before* he puts them in the receptacle, for example. A discussion on this type of cheating is given in chapter 10.

Placing the Bets—Maximums and Minimums

All players make their bets prior to the dealing of any cards for the hand. This bet cannot be changed once you have seen your cards, except in the cases of "insurance" and the exercising of certain player options. These exceptions are discussed in detail in ensuing passages. The casino assigns maximum and minimum bet levels for each blackjack table. The minimum may be as little as twenty-five cents and the maximum as high as $500. Nevada casinos generally offer the greatest range of maximum and minimum values.

Dealing the Cards

Once all bets have been placed the dealer makes one pass, from his left to his right, giving each player and himself one card, facedown.

On a second pass, all players are again dealt one card facedown, but the dealer's second card is turned up for all to see. In some places, like Aruba and Monte Carlo, the dealer will hold back his second card until he plays his hand. In these cases, his first card is dealt faceup.

The Natural or Blackjack

If your first two cards are an ace and a ten, or an ace and a face card, you have a "natural." A natural is more commonly referred to as a "blackjack" and we shall adopt this terminology hereafter. If you are dealt such a hand, turn your cards faceup immediately. You will then be paid 1.5 times your original bet unless the dealer also happens to have a blackjack. In that case the hand is considered a push and no money changes hands.

Insurance

Referring back two paragraphs, you will note that one of the dealer's cards is dealt faceup. If this card is an ace, he will immediately ask if anyone at the table wishes "insurance" and pause momentarily while you check your cards and decide.

Now, insurance is merely a *side bet* you can make that the dealer *does* have a blackjack. If you wish to make this bet, you place an amount equal to one-half, or less, of your original wager in the insurance area on the table. If the dealer *has* a blackjack, you will be paid *twice* the amount of your side bet. That is all there is to the mechanics, nothing more. There is no need to be frightened or confused by insurance as a great many people are.

Whether you take insurance or not, your original bet and hand are consummated in the normal manner. The insurance *side bet* has nothing whatsoever to do with the outcome of your hand and changes the play in no way.

Blackjack insurance is quite analogous to automobile insurance. The money invested is lost to the company (casino) unless, of course, you have an accident (a dealer blackjack). In these relatively rare instances, your losses are paid for by the insurance company. Just remember that all insurance companies are profitable businesses.

For playing my systems, I make a *blanket* recommendation *against* taking insurance—ever. However, intelligent and profitable insurance betting is possible for the "card counter." This type of situation is identified in chapter 9 on card counting for those interested readers.

Determining Hard and Soft Totals

You will recall that the *object* of the game is to obtain a higher *total* than the dealer without exceeding 21. If his total is higher than yours, or if you go over 21, you lose. If your total is greater than his, or he exceeds 21, you are paid even money.

In totaling your cards, all face cards count as 10, and all others count at their face or "pip" value. Aces may be counted as either 1 or 11 at your own option. A hand containing an ace is referred to as a "soft" hand if the ace may be counted as an 11 without the total exceeding 21. All other hands represent "hard" totals.

For example, an ace-seven combination is a soft 18. The addition of, say, a six will force you to count the ace as one, since you would exceed 21 otherwise. The ace-seven-six combination now becomes a hard 14.

To avoid confusion, first obtain a total, counting the ace as 11. If this total goes over 21, you have a hard hand so count the ace as a 1. If the total, tallying the ace as 11, does not exceed 21, then follow the recommended playing strategy for soft hands.

Drawing and Standing

After everyone has received and checked his cards, scrutinized the dealer's up card, and dispensed with the business of insurance (when applicable), additional cards may be drawn. Starting with the player at his extreme left, the dealer holds out the deck directly in front of him and sort of stares expectantly. The player will then indicate that he wishes another card, usually by scratching the surface of the felt with his cards, or make known that he will "stand" with what he has by sliding his cards facedown under his bet.

It is positively amazing how many variations there are in the way the people scratch their cards to ask for a "hit." The techniques range all the way from the short, quick snap to slow, deliberate dragging of the cards in repetitive motions. The latter is almost like slowly sweeping the floor. There *is* one thing that they all have in common —green felt under the fingernails. From this there can be no escape for the blackjack player.

In some casinos, usually in a two-deck game, the cards are all dealt faceup and you never touch them. Under these conditions, you wave your hand (think of all the possibilities here) or say "hit me," or something similar, when you want another card. Such a game has the virtue of producing players with the cleanest fingernails in the casino.

If the player desires a card, it will be dealt to him faceup. The dealer will then wait until the player indicates his intentions as before. He can draw as many cards as he wishes in this manner until he is either satisfied with his total, or it exceeds 21. Totals greater than 21 are termed "busts." If you go bust you lose automatically, then and there, and must announce this fact immediately by turning up your hole cards. The dealer will scoop up your money before proceeding to the next player.

You have the option of drawing or standing on any total whatsoever. The dealer does *not* have this privilege. He *must* draw to all totals of 16 or less and must stand on *all* totals of 17 or greater. He cannot draw to a hard, or soft, total that exceeds 17, regardless of what he thinks you have. The casinos in downtown Las Vegas require the dealer to hit a soft 17 as well. This play favors the house by 0.2 per cent.[2] The casinos on the Las Vegas Strip usually do not allow the dealer to hit a soft 17, however. Throughout all subsequent discussions, the downtown rules will be assumed unless otherwise noted.

This *fixed* strategy on the part of the dealer is very nice because it enables us to employ a *variable* strategy of our own that takes advantage of his inflexibility. The correct plan for drawing cards has been worked out by scientists employing electronic computers in conjunction with the laws of probability. Fortunately, the resultant strategems are not very difficult to learn and can be summarized with a few simple rules. This strategy is presented in complete detail in chapter 3.

Doubling Down

Upon receipt of your two cards you may, at your discretion, place them both faceup in front of your betting box and double the amount of your bet. You will then receive one, and only one, additional card, facedown. This player option is referred to as "doubling down" or, less frequently, as "going down for double."

Las Vegas rules most often allow doubling down on *any* two cards. Casinos in other locales generally restrict you to certain hard two-card totals only, such as 11, or perhaps 10 and 11. Some casinos that enforce these restrictions liberalize them slightly by allowing the specified totals to be achieved by *any* number of cards.

When correctly exercised, doubling down is a very powerful weapon for the player and can yield as much as a 1.73 per cent edge[3] over the player who does not utilize it. You are wasting your time and money playing in a game in which you cannot double down at all. Avoid such games as you would a roulette wheel with single, double and triple zeros.

As was the case for drawing and standing, complete precision strategies for doubling down on both hard and soft totals have been devised employing electronic computers and probability theory. Again, a few simple rules cover all possibilities. The doubling down strategy is given complete coverage in chapter 3. The dealer cannot double down.

Pair Splitting

Should your hole cards be any two of a kind, denoted a "pair" as in poker, you may double your bet and place them faceup in front of your money. Your original bet is associated with one of the cards and an equivalent amount with the other. Now you proceed to play *two* hands, each of which is built on one card of the original "split" pair.

Following the normal procedure, you may draw as many cards as you desire on each hand. In some cases, as in Aruba, though, you can draw only *one* more card on each hand. The one universal exception to this rule is for split aces. In this case, you get only one card on each ace. Should this be a 10 value card, the hand is counted as a plain 21, not a blackjack. There are no 1.5 to 1 payoffs on split aces.

Another possibility, especially in downtown Las Vegas, is doubling down after receiving a card on one or both members of the split pair. In succeeding chapters, this move is assumed valid unless otherwise specified. On the Strip in Las Vegas, and in some other casinos, doubling down is not allowed on a split. This restriction costs the player about 0.13 per cent.[4] The alert reader will notice an interesting fact about the two major rule differences between downtown Las Vegas and the Strip. Namely, downtown the dealer has the advantage of hitting a soft 17 but the player has the added option to double down on split pair elements. These two rule differences from the Strip basically cancel each other out percentagewise.

Finally, some casinos also allow further splitting of pairs after an initial split. This gambit gains anywhere from 0.53 per cent (1 deck) to 0.11 per cent (four decks).[5]

Just as for the other player options, there exists a precise computer developed strategy for splitting pairs. This is also detailed in chapter 3. The dealer cannot split pairs.

Folding Your Hand

This particular rule is very rare. In fact, I know of only two casinos, both in Las Vegas, where it may be found. It goes like this. If you do not like your first two cards, you may throw them in and pay the dealer one half the amount of your original bet. This terminates your play for that particular hand.

While folding your hand may appear at first to give the player more flexibility, it actually favors the house in most situations. For the basic strategy player, including those who play any of my systems, I recommend that you ignore this rule. If you are a card counter and are able to determine when the odds against you exceed 3 to 1, you may be able to gain by exercising this option. Otherwise, forget it.

SUMMARY OF THE DEALER'S RULES

As a quick reference for the novice, the dealer's rules are briefly summarized below. The dealer:
1. must draw to all 16s;
2. must stand on all 17s or greater;
3. must draw to a soft 17 (downtown Las Vegas, Reno, Lake Tahoe and Carson City);
4. cannot double down;
5. cannot split pairs;
6. does not win on a tie (push).

With the exception of rule 3, these rules apply in virtually *all* casinos offering the game.

These very stringent restrictions on the dealer will come as a welcome surprise to those of you who have played the home game but have never been in a casino. As you know, in the home game the dealer wins on all ties and can hit and stand as he pleases. For you who have played in the casino but are not familiar with the rules of the home game, refer to Hoyle.[6] You will find that the home game rules are disgustingly and hopelessly biased in favor of the dealer. Unless you can obtain that advantageous position, you do not stand a chance. You would be better off to save your money for the casino.

NOTES

1. George F. Hervey, *A Handbook of Card Games* (London: Paul Hamlyn, 1963), p. 284.

2. Edward O. Thorp, *Beat the Dealer* (New York: Vintage Books, 1966), p. 131.

3. Ibid., p. 131.

4. Ibid., p. 131.

5. Ibid., p. 131.

6. Albert H. Morehead, ed., *The Official Rules of Card Games* (Wisconsin: Whitman, 1959), pp. 228–231.

3. THE CORRECT BASIC STRATEGY

As you know, this entire book is devoted to the blackjack player who wishes to win consistently without remembering any cards that have been played previously. For such a player, there is only one *correct* way to play. That is, there is one, and only one, move for each combination of player's hand total and dealer's up card. This chapter is dedicated to teaching the complete fundamental strategy that encompasses all such situations.

My experience with friends and relatives has shown that the entire strategy can be learned in an hour or two of pleasant practice play at home. This learning process is easier and more pleasant still with a Du-Rite Wheel. For those of you willing to make this small effort, the rewards will be very great. You see, this method reduces the game to an essentially even one between you and the house! That is right! Learn the few simple rules presented below and you will have removed completely the usual casino edge. In fact, with a favorable set of house rules this strategy will actually produce a small (about 0.1 per cent) edge *in favor of the player*. Combining this method of play with the simple betting techniques disclosed in succeeding chapters will then yield player profit rates of 1 per cent and more.

To make things still easier for you, I have prepared a "reduced strategy" that has been summarized in a few simple rules. If you cannot learn them in ten or fifteen minutes, say en route to the casino, you might as well throw away this book and give up gambling. Despite the simplicity of the reduced strategy, it also produces an approximately even game. I estimate that this concise technique will cost you just a *small* fraction of 1 per cent. It too can win beautifully when combined with my new betting techniques.

DEVELOPMENT AND HISTORY

Mathematics has been applied to gambling for centuries. The first real attempt to employ some aspects of what is now termed "probability theory" was apparently made by a man named Gerolamo Cardano in the 1500s.[1] It is fairly well accepted that he was among the first to develop and employ some of the basic tools of modern probability. Although Cardano himself did not treat casino blackjack, which did not even exist in his day, probability theory has been applied extensively to the game.

Most of this effort has been accomplished in the last twenty-five years, primarily because of the availability of high-speed digital computers. Prior to the use of computers, the extensive amount of computations required to pin down the mathematically correct plays in all situations was quite prohibitive.

Nevertheless, an almost perfect basic strategy was published as early as September 1956, in the *Journal of the American Statistical Association*. The strategy was developed by four men (Baldwin, Cantey, Maisel and McDermott) at Aberdeen Proving Grounds, employing hand computations on ordinary desk calculators. For the record, it should be noted that they were "ably assisted by Otto Dysktra."[2] A great deal of credit should be accorded to these men, both for being the first to publish a strategy that very nearly eliminated the house edge and for exhibiting a monumental amount of perseverance in performing long-term hand computations. They state in their paper that their strategy reduced the house edge to –0.6 per cent. According to Wilson, a more correct value for their strategy would be –0.3 per cent.[3]

A number of individuals and groups have simulated this strategy, with very minor refinements, on large-scale digital computers. All simulations, amounting to many millions of hands of blackjack when taken in total, have produced the following result:

Fundamental, or basic strategy, blackjack is an essentially even game.

The player's edge obtained by computer runs made by various parties are given as follows[4]:

Date	Researcher	Computer Used	Player's Edge
1954	Atomic Energy Commission Los Alamos	IBM 701	–0.7
1956	Baldwin, et al. Aberdeen Proving Grounds	Hand Calculators	–0.3
1956	Ramo-Woolridge Los Angeles	Remington Rand 1103	–0.14
1959	Martin Denver	IBM 650	–0.01
1959	Univac, Los Angeles	Univac	–0.5
1960	General Dynamics Astronautics San Diego	IBM 650	+0.16
1961	MIT Cambridge	IBM 704	–0.21
1963	IBM Chicago	IBM 7044	+0.1
	To this I would add:		
1964– 1970	Dubey Cape Kennedy	Several	+0.13

The player's edge of about +0.1 per cent, derived by Julian Braun of the IBM Chicago Data-Center, is considered by experts to be the most valid. Employing a direct simulation of casino play, Braun has

run millions of hands on the computer employing the same strategy
that is presented in this chapter. It contains a very *few* refinements
over that strategy published by Baldwin, et al. in 1956.

As you can see, my multimillion-hand simulation is in excellent
agreement with Braun's results. My program duplicates the strategy
of this chapter in every detail. See chapter 4 for more about this
program, which is also a direct simulation of the casino game.

For those of you who do not unquestionably believe computers,
and being a mathematician and programmer myself, I can sympa-
thize with you, I have also run a ten-thousand-hand test of the Braun
strategy by hand. Casino regulations were adhered to rigorously and
all results were carefully recorded. These data yielded a +0.7 per cent
player edge for the Braun strategy. A similar ten-thousand-hand test
on an *inferior* playing strategy proposed by Goodman produced a
+0.4 per cent player's edge. While this sample size is not too significant statistically, it gives a clear *indication* of the strategy's validity.
If you still have doubts, run off a few hundred hands and get a feel
for it yourself.

THE REDUCED STRATEGY

The simple set of rules given below represents a close approximation to the complete basic strategy. You should be able to master
these rules very quickly. I would estimate that five minutes of memorizing followed by ten or fifteen minutes of practice should be more
than enough to lock in this simple plan. Yet you have lost very little
over the complete set of rules. The addition of the refinements to the
strategy presented subsequent to the reduced version will gain only a
small fraction of 1 per cent for the player. As previously noted, you
can use the reduced strategy in conjunction with either my simple
"Min-Max" or "Complete" betting systems (see chapter 5) and
become a winner.

1. *Drawing and Standing*

Hit all totals of 16 or less when the dealer shows a seven or
higher, otherwise stand when you have a stiff (hard 12 or
better). Stand on soft 18 or more.

2. **Hard Doubling**

Double down on all hard 11s, all hard 10s except when the dealer shows a nine or ten, and on hard 9s when the dealer shows a six or less.

3. **Soft Doubling**

Double down on all soft totals less than 19 when the dealer shows four, five, or six.

4. **Pair Splitting**

Always split aces and eights and never split fours, fives, or tens. Split all other pairs when the dealer shows seven or less.

5. **Insurance**

Never take insurance.

That is all! If you can remember those simple rules, and will bet the way I tell you later on, you will become an excruciating pinprick in the collective casino posterior. You will have the confidence and delightful profit that go with making the right play at the right time. The few minutes that it takes to learn these rules will be one of the best investments you have ever made if you intend to do any casino gambling at all. Otherwise, you are just stupidly throwing your money away. If you are too lazy to learn five simple rules, and you go and lose all your money at blackjack, you most certainly deserve it.

REFINEMENTS YIELDING THE FULL BASIC STRATEGY

With just a little more effort you can learn *all* the correct plays. The following additional rules will do it.

1. **Drawing and Standing**
   a) Hit a hard 12 when the dealer shows a two or a three.
   b) Stand on a 16 composed of three or more cards when the dealer shows a ten.
   c) Hit a soft 18 when the dealer shows a nine or a ten.
   d) Stand holding a pair of sevens against a dealer's ten.

2. **Hard Doubling**
   a) Double down on a hard 8 (unless it is composed of a six and a two) when the dealer shows a five or six.

3. *Soft Doubling*
   a) Double down on soft 17 and 18 when the dealer shows a three.
   b) Double down on soft 17 when the dealer shows a two.
4. *Pair Splitting*
   a) Split fours when the dealer shows a five.
   b) Split nines when the dealer shows nine or less (except when he shows a seven).
   c) Split sevens when the dealer shows eight or less.

The addition of these modifications to the reduced strategy will enable you to play the *best possible way* in all cases, assuming that you do not wish to keep track of the cards played. It is possible, but fairly difficult, to improve on this strategy if you are willing to keep track of exposed cards. However, it is the primary mission of this book to enable you to win without exerting yourself in this distasteful fashion. If you wish to know something about card counting anyway, please refer to chapter 10.

Appendix D contains a full explanation of the "Du-Rite" Strategy Wheel, which was devised to simplify this system even further. Appendix D also shows a representation of The "Du-Rite" Strategy Wheel, and provides the reader with information on how to obtain one.

SOME QUALITATIVE JUSTIFICATIONS FOR THE BASIC STRATEGY

Without delving into a lot of mathematics, I will give you a few of the intuitive and qualitative reasons that seem to justify some parts of the recommended approach. No attempt to be exhaustive with this section will be made. It is felt that this type of reading grows old very fast and rapidly becomes out-and-out boring. After all, the results already presented for millions of hands are really sufficient justification in themselves, are they not? All I want to accomplish here is to show that the technique advocated makes sense, at least from a gross analysis.

Drawing and Standing

Obviously, you must hit all hands that total 11 or less, since you cannot possibly go over 21 and any card must help your hand. It is also fairly obvious that you should stand if you have a 17 or higher and the dealer shows a rather bad card (two, three, four, five, six). You can at least tie the dealer if he also gets a 17. Moreover, since he quite possibly has a "breaking hand" or "stiff" (one in which he will exceed 21 if he gets a ten or picture card), he stands a good chance of going bust.

On the other hand, if the dealer's up card is a seven or higher, he may not have to draw any cards at all to make a 17. In this case, you lose the bust advantage. Also, you are holding a sure loser with your 12-16 if the dealer does, indeed, have a pat hand. Therefore, you should always hit your breaking hands of 16 or less when the dealer shows a seven or higher.

Now, as you know, an ace may be tallied as either a 1 or an 11. So the effect of what would be a poor hand, such as an ace and a five totaling 16, is "softened" by the fact that the hand may also be considered as a 6. Thus, some would-be breaking hands of 12-16 total, are often "soft" hands containing an ace and should be hit.

Soft hands of 12-17, as previously mentioned, must always be hit except when they qualify for doubling down. Why, you might ask, should you hit a soft 17? You could, you say, at least tie the dealer if he also has a 17. Well, the answer to this is that a 17 is a pretty bad hand to begin with since it can be beaten in a relatively large number of ways. Certainly you cannot hurt it much with a hit, and you might enhance it considerably with an ace, two, three, or four. That means that there are sixteen more cards (tens, jacks, queens, and kings) that will not change it. Thus, thirty-one cards out of a possible fifty remaining will either improve your soft 17 or maintain the status quo. Those are pretty fair odds. Furthermore, in many clubs the dealer will also hit a soft 17.

Note that for a soft, two-card 18, the helpful plus the noneffective cards drop in number to twenty-seven out of fifty. This is a slightly better than even chance, but you already have the dealer beaten if he has 17 and can tie him (push) if he has 18. It is also possible to turn your soft 18 into a bad breaking hand if you should get hit with a four, five, six, seven, or eight. There are twenty possible ways to get into trouble when you have already got a fair hand. Altogether, it should be clear that the situation in question is a marginal one, at best, for taking a hit.

On the other hand, if the dealer shows a nine or ten against your soft 18, he has a considerably higher chance of having a two-card 19 or 20 that will beat you. So hit.

For a soft 19, of course, the case for hitting becomes pretty grim. Here you can only help the hand with seven cards, three aces, or four deuces, and can hurt it with any of twenty-seven cards (3–9). Hitting a soft 19 is a very poor play, then, and should be avoided.

Hard Doubling

Doubling down on two-card totals of 9, 10, and 11 will more often than not produce a high final total. The ratio of cards that will produce 17 or better to those that will not are given below. The dealer's cards are not considered in the following ratios.

Player's Two-Card Total	Ratios of Possible High Hands to Possible Low Hands
9	28/22
10	32/18 or 31/19
11	32/18 or 31/19

You can see that you have to be a little more careful and make good use of what the dealer shows when your total proceeds downward from 10. This, of course, is exactly what the strategy recommends.

Soft Doubling

On any soft hand, you can hit without going bust. Thus, doubling down here would not ever send your total over 21. Moreover, many soft totals can be improved or unchanged by a majority of the remaining cards. This can be seen by use of the same type of reasoning given above for betting soft hands. I leave this as an exercise for the interested reader. Note that the worst cards a dealer can have, for his sake, are the four, five, and six. These are up cards that predominate in the soft doubling strategy.

Splitting Pairs

If you get a pair of deuces or threes, they should always be split and played as two hands if the dealer shows 7 or less. Together they make bad starting hands of 4 or 6, respectively, and as such, are likely to become the very bad breaking hands of 14 and 16. When split, however, they are not quite as disastrous if hit with a ten or picture card. Then too, you now have two chances to draw a decent hand. If one of them turns out to be a winner, then you have turned what was very likely to be a losing hand into a draw, moneywise. Since the dealer must also be prone to having a poor hand (he must show 7 or less) when you split twos and threes, you have a better chance of at least *breaking even* with the two hands. Similar reasoning applies to splitting sevens and eights since these already total a miserable 14 and 16, respectively.

Aces should always be split. Together they total only 2 or 12. When split they are very nice 11s. There are twenty-eight cards that can be drawn that will give you a high hand total of 18 or better. These are sevens, eights, nines, tens, jacks, queens, and kings. So even though you will only get one additional card on each, you still have a better than even (twenty-eight out of fifty cards) chance to get a sticking hand. So split those aces, by all means.

Now, pairs of fours and fives total 8 and 10, respectively. When split they make very bad starting hands. Why split one good hand into two bad hands?

Incidentally, although I do not want to frighten the laymen among you, when you figure the ratio of good to bad cards remaining in the full, unseen deck as we have been doing, you are actually examining the "probability" of getting a good card next. Unwittingly, you have been using probability theory to justify the basic strategy all along. You rascal, you!

THE BASIC STRATEGY FOR FOUR DECKS

There is a basic strategy for four decks devised by Julian Braun of IBM. This strategy contains the following changes from the one-deck strategy as extracted from Revere's tables:[5]

1. Do not double down on five, three against a dealer's up card of five or six.
2. Do not double down on hard 9 against a two up.
3. Do not double down on hard 11 against an ace up.
4. Do not double down on ace, two or ace, three against a four up.
5. Do not double down on ace, six against a two up.
6. Do not split a pair of sevens against an eight up.
7. Split a pair of fours against a six up.
8. Hit a pair of sevens against a ten up.
9. Hit ace, seven against an ace up.

None of these changes is of tremendous consequence, but they should be implemented if you wish to play precisely in a four-deck game.

NOTES

1. Gerolamo Cardano, *The Book on Games of Chance,* trans. Sydney Henry Gould (New York: Holt, Rinehart and Winston, 1961), pp. 5–55.

2. The quotation is taken from a footnote to their article. *Journal of the American Statistical Association,* 51 (1956), pp. 429–439.

3. Allan N. Wilson, *The Casino Gamblers' Guide* (New York: Harper and Row, 1965), p. 92.

4. Ibid., p. 91.

5. Lawrence Revere, *Playing Blackjack as a Business* (Las Vegas: Paul Mann, 1969), p. 29.

4. A NEW COMPUTER APPROACH

As we have seen in chapter 3, computers have been slashing away at blackjack for years. So far, these attacks have produced a "basic strategy" (chapter 3), yielding an approximately even game and numerous successful betting techniques based on remembering cards or "counting points" as they fall.

My approach utilizes a direct simulation of the basic strategy to yield statistics from which a simple winning betting scheme may be constructed. The beauty of the new scheme, as you have been promised, is that it requires no counting of points or remembering cards.

A NEW CONCEPT

The key to the new approach is the type of statistics kept. Every completed hand is assigned to a category. For reasons that will be disclosed later, I used the following classifications:

a) Splits (not including split aces)
b) Split aces
c) Hard double-downs
d) Soft double-downs
e) Four-card hands (completed using a total of four cards including the player's and dealer's hands).
f) Pushes (excluding four-card hands).
g) Ordinary wins (excluding all of the above)
h) Ordinary losses (excluding all of the above)
i) Seven-card hands (completed in seven or more cards).

Now here is the critical step. For each of these categories, the computer program keeps track of the total number of units won and lost *on the next hand.*

For example, in the split pair category, the statistics "total units lost after splits" and "total units won after splits" are accumulated. Similar statistics are tallied for all of the other classes. If we analyze a very large number of hands, these data will enable us to determine whether our bet size should be increased or decreased after completing a particular type of hand. Obviously, if we consistently show a significant tendency to win more units than we lose after, say, split pairs, then it will be profitable to raise our bet on all immediate successors of this type of hand; especially if we have expected this all along and can rationalize the results achieved. Conversely, we should clearly bet the table minimum after a hand type that tends to be followed by more losses than wins.

In this manner, it is possible to devise a *simple* betting method based on the *type* of the last hand *without regard for the individual cards.* Thus, we have a winning betting scheme that does not require keeping track of the cards or counting up assigned points. Nor does it require a variable playing strategy.

It is possible, by examining the nature of the basic strategy, to predict what will happen in each category without a computer, at least qualitatively. This type of prediction presumes an elementary knowledge of card counting theory that is easily grasped. A pertinent discussion of the needed concepts is given where needed in the ensuing system descriptions and accompanying justifications.

The actual results in all of the above categories are presented later on in this chapter. In chapter 5, the systems constructed from this data are detailed along with discussions on what was expected in each category and why these particular classifications were chosen in the first place.

To compile the statistics needed to construct a valid system accurately, it was necessary to examine the results of millions of hands. The computer was far better equipped to do this than a person. Read on to find out how much better.

MAN VERSUS MACHINE

As dim-witted as it is, the modern-day digital computer is a fantastic tool. Once it is told (programmed) precisely what to do, it will grind out perfect answers with mind-boggling speed. For example, my program is actually somewhat inefficient from a running-speed standpoint. Yet, the Univac 1230 computer uses it to play ten thousand hands of perfect "basic strategy" blackjack in just about five minutes! Better still, the XDS Sigma 7 can accomplish the same task in twenty-four seconds! This includes a lengthy shuffle routine employing three complete interlacings of the cards (about one shuffle every nine hands), keeping track of several betting systems and a host of statistics, and listing the results on a high-speed printer.

You can stand by the compute module of the slower Univac 1230 and watch the lights blink in an endless pattern when the blackjack program is running. There are clearly some four major blinks per second. I have hypothesized that each one of these primary blinks represents a new shuffle of the deck! This theory yields about the right number of hands per minute as observed on the printer. So, at this sluggish two thousand-hand-per-minute clip (electronically speaking), the computer is accomplishing each minute what would take two people about eight hours—and the computer does not goof.

Looking at it another way, the computer inefficiently produced five million hands, plus accumulated statistics, in a matter of hours.

It would take two men, working forty hours per week, an estimated ten years to accomplish this herculean task!

The use of a still faster computer, along with a souped-up program, could increase the output rate considerably. Nevertheless, as so succinctly put by Wilson,[1] the current program "satisfied my aesthetic needs".

For anyone interested in pursuing the study of blackjack on a digital computer, a brief description of the organization and design of my program is next given. This outline may serve as a springboard to your own model.

THE PROGRAM

General Description

My blackjack program represents a "direct simulation" of the casino game. That is, every step of the actual process of casino play is duplicated by the computer. Nothing is left out.

The cards are dealt into the player's and dealer's hands, one at a time. The player completes his hand in precision accordance with the basic strategy of chapter 3. The dealer completes his hand, after the player, employing strict adherence to the typical casino rules of chapter 2. Only a single player has been built into the program, but you will see that this has no appreciable effect on the validity of the results, or the systems constructed from them. Splitting of a single pair is simulated, but no further splitting is allowed on a given hand. Doubling down on soft and hard totals is included, even on split pairs.

When fifty-one cards of the fifty-two-card deck have been played, the deck is shuffled using pseudorandom-cut depths and triple interlacing. The shuffling method has been designed to be realistic and to make long-term cycling impossible. In shuffling, a card is burned before normal play is resumed. If cards are "on the table" (incompleted hand) when the deck is exhausted, they are left there and the remaining cards are shuffled. Play then continues from the remaining cards, picking up the hand that was in play at the time of the shuffle and completing it.

At the conclusion of each hand, all pertinent statistics are updated and stored in the computer's memory. Cumulative profits and percentages are also computed for the basic strategy and versions of the "Min-Max" scheme and "Complete System" described in chapter 5.

All data are listed, on-line, at an optional print interval. For program checkout, the data were listed after each hand. For the long runs of a million or more hands, a ten-thousand-hand print interval was selected. For fluctuation studies, an interval of 100 hands was also used.

Blackjack Program Organization

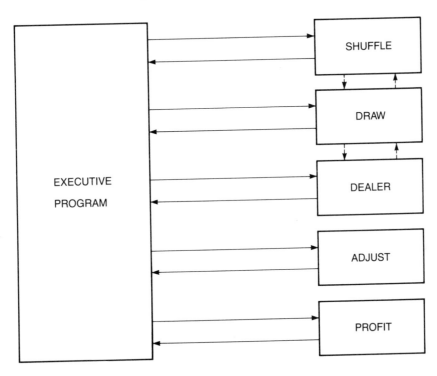

Organization

There are as many ways to design and organize a program of this magnitude as there are programmer-analysts. You will no doubt think of a method of doing the job that seems better, or has more appeal to you. Such is life. My scheme works beautifully and can satiate adequately my lust for correct data. Should you, for some reason, find this method totally horrifying, please communicate your suggested improvements to me for my edification.

As you can see from the diagram, the program consists of a main program, or "executive," and five subroutines.

Executive Program

The exec embodies the overall program logic and calls upon the subroutines as needed. It contains the complete basic strategy decision matrix and controls the dealing of the cards, playing of the hands and all other housekeeping tasks.

Subroutines

SUBROUTINE DRAW. With the exception of the first two cards given to both the player and the dealer, a new card is obtained by calling this subroutine. That is, any hand is "hit," be it the player's or the dealer's, by calling on DRAW. If the deck has run out, DRAW will call upon subroutine SHUFFLE to shuffle the cards. The appropriate hand is then hit from the new deck and control is returned to the main program. This routine may also be called by subroutine DEALER which is described next.

SUBROUTINE DEALER. The purpose of DEALER is to play the dealer's hand. Having accomplished this, DEALER compares the player's and dealer's final totals, assigns a code indicating whether the hand was a win, loss, or draw, and returns control to the main program. Since the dealer's hand must often be hit, subroutine

DRAW may be utilized by DEALER. DRAW may in turn call upon SHUFFLE as needed.

SUBROUTINE ADJUST. ADJUST is called by the exec at the conclusion of each hand. Using a code assigned by the main program that identifies the "type" of the previous hand (split, double-down, ordinary win, etc.), the appropriate statistic is adjusted in accordance with the outcome of the current hand. Control is then regained by the main program.

SUBROUTINE PROFIT. System profits for built-in bet values are computed by this subroutine at the completion of all hands except pushes. Desirable statistics other than those accumulated by ADJUST are also maintained by PROFIT. This routine interfaces with the main program only.

SUBROUTINE SHUFFLE. A great deal of thought was given to the design of SHUFFLE, since it had to be beyond reproach if the simulation results were to be valid. I also wanted to develop something unique to provide still another computer verification of the basic strategy edge; especially since I am promoting the Du-Rite Wheel (see chapter 3).

Moreover, as close an approximation to the actual single-deck game as possible was desired. That is, pseudorandom cutting biased toward the center of the deck, thorough card interlacing in a systematic fashion, burning of a card, and the leaving of the in-play cards on the table.

Please do not say I should have used a random number generator to simulate random draw—with or without replacement. That is what everybody else has done. Yes, it is simpler, but in my book it is not as true a simulation as is obtained with SHUFFLE. Besides, it is always of extra value and interest when an *independent* scientific investigation produces good results through a heretofore unexplored avenue.

Basically, SHUFFLE cuts the deck three times. Following each cut, the deck is split and the cards are thoroughly interlaced. The first cut depth is calculated by an algebraic function of a pseudorandom number. This is a code number corresponding to the particular "type" of the previous hand.[2] The second cut depth is increased by one card every forty thousand hands with a cycle of about two million hands. The final cut depth is increased by one card every two thousand hands with a cycle of forty-eight thousand hands.

An analysis of the phasing of these cut depths will convince you that long-term cycling (playing a particular set of hands over and over again) is impossible with this routine. Also, the first cut depth is very random in nature. This, together with the interlacing, produces a very satisfying mix indeed.

Program Language

Since it was not known on which computer I would be running the program when I wrote it, only basic FORTRAN was used. Because of this, the program can be quickly compiled and executed on most any of today's high speed, general purpose, digital computers.

Unless you wish your program to be tied to a specific computer, I would steer clear of the more "efficient" machine language approach. Otherwise, your program may die an unnatural, and unnecessarily early, death when you suddenly find you no longer have access to the original computer. Use FORTRAN, or a similar high level, universal language and spare yourself the agony of converting your work at a later date.

As an aside, the lack of this type of foresight has cost the U.S. government, as well as industry, many, many megabucks for program conversion at computer changeover time. It is a lesson to be well learned by all software people, with emphasis on the assembly language addicts. What is saved by machine language efficiency may be thoroughly inundated by ensuing conversion costs.

SIMULATION RESULTS

The name of the game is to discover what happens *after* each "type" of hand, as we have previously categorized them. That is, what is the expected percentage in favor of the player, or against him, on the hand that follows?

The table that follows gives the total units won and lost, and the percentage edge *after* each type of hand. A plus (+) sign indicates an edge for the player and a minus (–) sign indicates an edge for the house. The "percentage after" indicates what would happen if you flat bet all hands immediately following those in any given category. Results are based upon the data from computer runs totaling five million hands of basic strategy blackjack.

Type Hand*	Units Won on Following Hand	Units Lost on Following Hand	Percentage After
SPLITS (not aces)	40,684.0	39,564.0	+1.4
SPLIT ACES	11,389.0	11,777.0	–1.7
HARD DOUBLE-DOWNS	227,586.0	222,165.0	+1.2
SOFT DOUBLE-DOWNS	51,037.0	50,580.0	+0.5
FOUR-CARD HANDS	465,273.0	475,477.0	–1.1
PUSHES (not four-card)	139,465.5	138,061.0	+0.5
ALL SEVEN-OR-MORE-CARD HANDS	367,305.5	360,099.0	+1.0
ORDINARY WINS	769,623.0	767,122.0	+0.2
ORDINARY LOSSES	937,382.5	930,757.0	+0.4

*Full definitions of these categories follow (immediately after this table).

In the *SPLITS* category, split aces are not included. All other hands which called for pair splitting using the basic strategy are included.

The *SPLIT ACES* category is self-explanatory. Every pair of aces is split, of course, as we know from chapter 3.

Again, the *HARD* and *SOFT DOUBLE-DOWN* categories include all hands of these types as defined by the basic strategy.

The *FOUR-CARD HAND* category includes all hands won, lost, or tied, that were completed using a combined total of four cards only. That is, two cards for the player, and two cards for the dealer.

A *PUSH* is any hand where the dealer and player wound up with equivalent totals that were less than or equal to 21. However, four-card pushes were tallied in the *FOUR-CARD HAND* category and were not counted as pushes here.

The *SEVEN-CARD HAND* category includes all hands completed using a joint player-dealer total of *at least* seven cards. Now, since all *SEVEN-CARD HANDS* qualify for other categories as well, they were also tallied there. For example, a split hand using a total of eight cards would be included in both the *SPLIT HAND* and *SEVEN-CARD HAND* statistics. In other words, all *SEVEN-CARD HAND* statistics given in the preceding table are redundant.

ORDINARY WINS are any winning hands that do not otherwise qualify for a special category (excluding the *SEVEN-CARD HAND* category). Likewise for *ORDINARY LOSSES*.

In the next chapter, we will make use of these data to construct our new betting systems.

NOTES

1. Wilson, *The Casino Gambler's Guide,* (New York: Harper & Row, 1965), p. 81.

2. As a control, multimillion-hand runs were also made employing a first cut depth obtained from a true, random number generator with a cycle in excess of one hundred thousand hands. All three cut depths are phased so that the total "cutting cycle," if you will, is many millions of hands. Even at this point, duplication of previous hands would be extremely unlikely because of the feature of leaving in play cards on the table and out of the shuffle. In any event, multimillion-hand runs with either the random or pseudorandom cut depths gave essentially identical results, statistically.

5. DUBEY'S METHOD

In this chapter, you will get what you have been promised—a winning blackjack betting method that requires no perceptible mental effort on your part. That is right! You can gain up to a 1 per cent edge on the house without employing either card counting or a variable strategy.

For my betting techniques to work, you need only play according to the fundamental strategy of chapter 3 and bet the way I will tell you. Even my simple "reduced strategy" will be adequate for short-term gambling, although the addition of the refinements is highly desirable.

First you will be introduced to my "Min-Max" scheme, which is outrageously simple, yet yields up to a 0.3 per cent edge for the player. Finally, my "Complete System" is presented, which is almost as easy. It will get you as high as a 1 per cent edge on the house; again without remembering or counting a single point.

A PRECAUTIONARY NOTE

Before you go to the casinos, you must decide how much you can afford to gamble, assuming that the worst will happen and you will lose your stake. This is vital to winning. You must never gamble with money that makes you sweat! You cannot be afraid, at the critical moment, to make that appropriate large-size bet. Also, since your edge on the house will not be exactly tremendous (on the order of 1 per cent), your financial status will be subject to what are often rather frightening fluctuations. Despite the care with which you choose your bet sizes, there will always remain a small probability that you will be wiped out.

Once you have decided how much you can afford to lose without crying about it, you are ready to select the size of your *maximum* bet. To determine what this maximum should be, and what the associated chance of ruin is, you should consult chapter 6.

However, as a rule of thumb, your stake should equal at least 100 to 150 times your maximum bet. If you obey this rule, you will have a good chance of doubling your money without losing your stake. Please note that about ten people out of every 100 would actually lose their bankroll if they all obeyed this rule. The other ninety, on the other hand, would successfully double their money.

As I said, if you want to know exactly what is required to reduce this chance of ruin still further, you should consult the following chapter.

THE MIN-MAX SYSTEM[1]

Okay, now for my Min-Max betting system. First, your blackjack playing strategy must conform to that of chapter 3 for the system to be effective. Get out your Du-Rite Wheel and practice for an hour, or at least take ten minutes to memorize the "reduced strategy" (see chapter 3). Learning the fundamental game is requisite to winning at anything, and blackjack is no exception.

Having accomplished this, you need only bet as follows:

After all losses and all four-card hands, bet the table minimum. Otherwise bet your maximum.

Now, I ask you, what could be simpler than that? Yet, as amazing as it seems, this simple system can produce an edge for the player which is some three times higher than the 0.1 per cent obtained by the basic strategy of chapter 3.

Note that this approach has you betting at *your* maximum level after all wins and pushes, unless they happen to be four-card hands.

It does not at all matter where you sit, nor is it particularly important how many players there are at the table. It is best to pick a table with as few players as possible. This is because you lose your advantage on the next hand when the deck is shuffled, and it will be shuffled more frequently if the table is crowded. You can use a crowded table to your advantage also, however, if you practice a little "leaning" as described at the end of this chapter.

As for multideck games, avoid them if possible. The greater number of cards will tend to dilute your advantage although it will still exist to a lesser degree. Most Las Vegas houses still offer plenty of single-deck games, especially downtown.

Why Does Min-Max Work?

The quick answer to this question is easy to find. As we saw in chapter 4, my computer results showed that flat betting all hands immediately following four-card hands would yield –1.1 per cent for the player. That is, the house has an average advantage of about 1.1 per cent just after a four-card hand.

This being the case, you can negate some of the house advantage, and *thereby gain some edge for yourself,* by simply betting the minimum allowable after every four-card hand.

In like fashion, the computer showed that there is about a 0.5 per cent flat bet advantage for the player on hands that immediately follow a push. Therefore, you can pick up some more personal advantage by consistently increasing your bet after a push, provided it was not a four-card hand.

As for increasing the bet after wins, and betting the minimum after all losses, this is more a matter of good money management than anything else. The computer results showed only a very slight bias for the player after wins. It did, however, indicate that there may be a little more significant player's edge after losses once the four-card hands have been excluded. Wilson also hinted at this (without the four-card exclusion) in his book.[2] Personally, I am skeptical about the value of this particular bias, whether it exists or not, from a practical standpoint.

With the Min-Max approach, good money management is automatic. If you have a long series of losses, they will be at the table minimum betting level. If you have a long run of wins, on the other hand, they will essentially all be at your maximum bet level. Later on, I will tell you about some extensive casino workouts of this scheme that really proved it out in action.

By now, you are no doubt wondering what this "four-card" business is all about. You are probably thinking something like "Just what in the name of Thorp has a four-card hand, or a push for that matter, got to do with what happens on the next hand?" Moreover, you may at this point be unimpressed just because a computer played a few million hands and came up with some numbers. Right?

Well, I do not blame you, so we will try to rectify that situation now. While the explanations here could be very complex, I will attempt to put the supporting evidence to you in simple layman's terms on the premise that most of you are not mathematicians or statistical analysts. I want to *teach* you, not *snow* you.

First of all, you can easily understand that blackjack is very different in nature from most other gambling games. Clearly, in craps, for example, one toss of the dice has absolutely no bearing on the results of the succeeding toss. The dice have no memory. They could not care less what happened on the last toss, no matter how much money you have riding, or how many times in a row the number seven has come up.

In blackjack, though, we deal with a deck of cards. Some of these cards, like aces and tens, are favorable to the player when present in the deck. Conversely, other cards like fives and sixes affect the player adversely when they predominate in the deck. This is why Dr. Thorp and other experts have been so successful in devising and exploiting point-count systems that determine when the deck is rich or poor in good cards. When, through a priori knowledge of what has been played, they find that the remaining deck has become favorable to the player, they raise their betting level. Taking advantage of the fact that succeeding outcomes will be heavily dependent on what has been played, then, they gain an edge over the house.

Well, Min-Max betting is also based upon exploiting this fact that wins and losses do not occur in perfectly random fashion. This is because certain types of hands are very likely to use up an excess of bad cards or good cards. If we can gain an understanding of what kind of hands use up what kind of cards, we can then predict what will happen, on the average, on the next hand.

Before we go any further with this discussion, let us take a gross look at which cards are considered good for you, and which are bad when they remain in the deck. It has been shown mathematically that the general effect *on the player* of an excess of various cards left in the deck is as follows:

Good Cards	Ace, 10, J, Q, K.
Bad Cards	2, 3, 4, 5, 6.
Cards with Little Effect	7, 8, 9.

So we can now generalize, if a certain type of hand tends to use up a lot of *low* cards, then it would also tend to be followed by a win. Not *every* time, you understand, but more often than not. Conversely, a type of hand that consistently used up high cards would have a favorable probability of being followed by a loss.

This leads us right back to our four-card hand. Remember, this hand is completed with just two cards each for the player and dealer. Now, you know that the dealer must have a total of 17 or more to stand. Thus he must have at least one high card, and no low cards, to stand with just two cards. The same thing applies to the player. The basic strategy says to hit if you have a low total and the dealer, as he must in this case, shows a seven or better. So, all four-card hands will use an excess of high cards; a situation that leaves the deck less favorable to the player. It is because of this that the computer got the results it did and the Min-Max rule for four-card hands works.

That the push tends to leave the deck more favorable to the player, excluding four-card pushes, is less obvious. Pushes, by definition, require equal player-dealer totals between 17 and 21. These high totals can be achieved by either a few high cards or a lot of low cards. It can be demonstrated by examining the probabilities that the push, on the average, will use up an excess of low cards.

Probabilities aside, though, it is easy to see that a push consisting of six or more cards must certainly average more low than high cards. One high and two low cards for each hand would be typical for a six-card push, for example. Aces counted as one do have a natural dampening effect, of course, but there are only four of them, remember.

The five-card push is particularly questionable as a user-up of low cards. If you wish, you could add a refinement to the Min-Max scheme that would exclude raising your bet after five-card pushes; but this requires recognizing them and really is not worth it, to my way of thinking. You cannot miss a four-card hand because of the absence of hitting. No work is required. But a five-card hand requires a little counting effort, and this is definitely a no-no for readers of this book.

Min-Max in the Casino

Together with George and Mary Sykes, I have tested a very conservative version of the Min-Max system in most of the Las Vegas casinos. Some six thousand hands were played on two consecutive weekends in this fabulous town.

A two-dollar maximum and one-dollar minimum was used exclusively for the test to minimize both our stake size and the anticipated casino reaction. We did not, however, exclude four-card hands for this test, but went ahead and bet the two dollars after all wins and pushes. As you know, this is a very conservative approach to Min-Max betting indeed! Without the four-card rule, however, our system in no way resembled card counting. We felt, and justifiably so, that our winning would be tolerated a lot longer by the dealers this way. We were right, as you shall see.

George (a favorite uncle) and I played approximately 2,600 hands on the first weekend. The total of 2,600 hands is an estimate, but a fairly close one. For each session of play we recorded, among other things, the starting time, the ending time, and the average number of players at the table. By clocking the rates at which dealers typically cranked out hands as a function of the number of players, we concluded that we should estimate anywhere from 50 to 120 hands per hour, with the total increasing with decreasing numbers of players. A certain amount of judgment was required for each session, but since these were rarely in excess of an hour in duration, we could not get too far off. Our estimates are most likely a bit conservative.

The same procedure was followed for the second weekend, yielding a sum total of approximately six thousand hands. As you can imagine, we were kept pretty busy achieving this amount, playing around the clock except for meal breaks, one show and rare two-hour naps. We had a higher hand total for the second weekend, since this time we were accompanied by my Aunt Mary.

Well, we got off to a great start! I first sat down to a single-deck game in the Fremont with one other player at the table, while George was busy making reservations for us to see the Giselle MacKenzie show that evening. When he arrived at the table a half hour later, he asked, "How're you doing?" "Winning about fifty bucks," I answered joyously. Up to now he had been a little skeptical about our prospects, but he sure watched intently for the next half hour.

An hour had gone by since I had started playing and in keeping with what was to become our general practice of playing no longer than one hour at a table, I got up and cashed in my chips. I was an amazing $42.50 ahead for my first hour of play! First blood had been drawn on what was to prove one of the best of several such relatively heavy winning streaks.

Our financial progress is depicted in a graph with winnings plotted versus the estimated number of hands played. If you follow the solid line through to 2,600 hands, you will see that we wound up winning $185 at that point. This corresponds to the end of the first weekend.

Although we occasionally did switch to a three-dollar maximum bet for very brief periods, the great bulk of our betting, say 99 per cent, was at the two dollar level. You can get an idea of how pleasantly surprised we were with the results by examining the other line in the figure that corresponds to the predicted profit for a 0.2 per cent edge. Even this would be somewhat optimistic for the bet level and system employed for this test. We had not as yet experienced any appreciable losing streaks. Consequently, our winnings just continued to mount up, right through the weekend, leaving us way ahead of the predictions at this point.

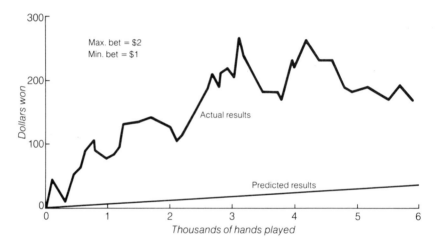

Casino Results
Min-Max System

We found that hot streaks were the backbone of our winnings, just as occasional cold sessions constituted the great bulk of our losses. For the most part, we would play long periods when we just held our own, never pulling very far ahead or falling very much behind. Then, in sudden rushes, our fortune would change considerably in just a few minutes. The most dramatic changes were in the winning direction, however. This is characteristic of Min-Max betting. When you win a lot, you win at your maximum bet level. During losing runs conversely, you lose at your minimum bet level. Such is the beauty of the Min-Max scheme.

If you once again refer to the graph and trace out our record of the second weekend, 2,600 to 5,900 hands, you will notice that we did not fare nearly so well this time. In fact, we ended up in the red by $12.50 for the entire weekend of 3,300 hands. This reduced our winnings to $172.50 for the combined two-weekend test. All in all, the results were very gratifying. Not only did we win, we far exceeded our predictions.

I attribute our outstanding first weekend's success to a good system and a good deal of luck (a positive fluctuation). The break-even second weekend was really better than it might appear. For one thing, quite a few costly playing errors were made when we got exhausted after many hours at the tables. For another, George and I were now well known by the dealers and were subjected to ever more frequent casino countermeasures as the second weekend wore on. See chapter 10 for the details and specific incidents.

Aside from the increasingly aggravating trials of blackjack, we experienced a number of interesting incidents typical of the fabulous Las Vegas. During one lengthy session at a table located near the slot machines, we were interrupted by the voice of a woman. "Isn't that Liberace over there?" she asked. I looked over my shoulder down the nearest row of one-armed bandits and sure enough, there was the famous Liberace. Not ten feet away, he was whaling away at the slot machines and grinning broadly. He was a spectacular sight with his snappy attire, tremendous head of hair, ever-present smile and flailing arms. The man sitting next to me, a very serious, dead-pan Southerner, summed it all up nicely as he slowly drawled, "Wild looking stud, ain't he?"

Yes, Vegas is a fascinating town and you will love it, I am sure. But the town will never be as fond of you, rest assured, if you play black-jack and use my "Complete System," which is described next.

THE COMPLETE SYSTEM

Actually, this system is only slightly more difficult than the Min-Max approach. It will indeed raise your edge to about 1 per cent at the high maximum bet levels and provides a lot of excitement and action. Carried to the limit, which I do not recommend you try, it will produce a maximum mean profit rate of about $120 per hour.

Here is how it works:

1. **Bet the** *table* **minimum after all:**
   four-card hands
   ordinary wins
   ordinary losses
   split aces
2. **Bet** *your* **maximum after all:**
   double downs
   splits
   seven-card hands
   pushes

That is the whole system! All you need to know is the "type" of the last hand to determine whether your next bet will be at your maximum or minimum level. How can you miss that?

Again, you need only play the basic strategy or, less desirably, my "reduced strategy." No strategy variations will be required. The remarks made for the Min-Max system pertaining to seat position, number of players, shuffling and multideck games are equally applicable to the Complete System.

Note that rule one prohibits raising the bet after all four-card hands. This means that four-card pushes are excluded in rule two. Also, remember that a seven-card hand, by my definition, is any hand that uses up seven or more cards between you and the dealer.

Why Does the Complete System Work?

The empirical proof, based upon a computer test of five million hands, is summarized in the table in chapter 4. This table shows that the *house* has the edge after those types of hands specified by rule one, above. On hands following those in the categories identified in rule two, the table shows that *you* have the edge.

Thus, following this system will produce the *maximum profit rate* for your particular upper limit on bet size. You can determine what this should be from the chapter on stake size and fluctuations.

Of course, a twenty thousand-hand home test run off by hand and our documented casino results also provide additional empirical support for the system. However, a little discussion here will give you sound logical arguments that predict the observed results.

To this end, four-card hands, ordinary wins and losses, and pushes have already been covered in the Min-Max discussions earlier. Significantly, a minimum bet is employed after ordinary losses for purposes of good money management.

That soft double downs do not heavily contribute to an appreciable bias in the deck in either direction is fairly obvious. An ace and a low card, plus one variable card, are always used by the player. Only a small advantage here. The dealer will be showing a low card —true—but his remaining cards could be anything. As the table in chapter 4 shows, the computer got +0.5 per cent (a player's advantage) based on the immediate successors of some one hundred thousand soft doubles.

Split aces always remove two good cards (the aces themselves) from the deck, although all other cards in the hand will be essentially distributed randomly, valuewise. Our prediction here would then be for a significant bias in favor of the house on hands following a pair of split aces. The computer got –1.7 per cent (a house advantage) based on approximately twenty thousand sets of split aces. This is a particularly small sample and the magnitude, but not the sign, of this result is somewhat suspect.

Next, take a close look at the basic strategy for hard double-downs. You will notice two interesting features of this strategy: (1) you only double down on hard two-card totals of 8, 9, 10, or 11; (2) in *most* cases, the dealer will be showing a low card when you double down.

Also, please appreciate the fact that it will almost always take two low cards, or a low and a noneffective one, to get you a hard total of

8, 9, 10, or 11. This means that you most often see *three* low cards when you execute a hard double-down. An assumption that the remaining cards in the hand will be about equally divided between good and bad cards is also reasonable. (Actually, a very slight excess of high cards may be postulated for those completing the hands on the average.)

Our prediction, based on the forgoing expected usage of more low than high cards, would definitely be in favor of the *player* on hands immediately following a hard double-down. The computer got a +1.2 per cent player's edge following some four hundred fifty thousand hard double-downs.

Examining the pair-splitting strategy (excluding aces) is equally enlightening. You only split low or noneffective valued pairs. Once again, you will almost invariably have to see a low dealer's up card before you will split your pair. The effect and prediction here is very like the hard-doubling situation. That is, we would expect a player's edge on hands immediately following a split pair. The computer got a +1.4 per cent edge based on some eighty thousand hands following split pairs.

Finally, consider the hand completed in seven or more cards. The maximum amount of high cards that could be used, excluding aces in the "soft" sense for the moment, would be four in any given hand. I submit that this eventuality is highly unlikely for a hand using seven or more cards. Clearly, such a hand will most often contain an excess of low cards.

The ace counted as 1 is a dampening factor here unquestionably. Since there are four aces and approximately nine hands per fifty-two-card deck (assuming only one player), an ace will be a factor in a maximum of roughly 40 per cent of all hands. Looking at it another way, about 60 per cent of our seven-card hands will not be affected by aces. While aces will not blow our theory apart for seven-card hands, then, they will certainly weaken it.

All in all, we would expect a player's edge after this type of hand since an excess of low cards will normally be employed in its average makeup. The computer got +1.0 per cent for the player following an approximate total of three hundred seventy-five thousand seven-or-more-card hands.

It should be noted that these remarks about soft aces also apply to the push discussion earlier in this chapter in the Min-Max section.

This completes our discussion and qualitative analysis of all elements of the method. It is possible, if overwhelmingly arduous, to arrive at these predictions from direct probability calculations. For you probability fiends, you might devote a substantial portion of your life to such an effort and have a great time for yourself. I would be very appreciative of your results, naturally.

Some Other Questions

A very good question you could raise might be, "Why confine your findings to only the hand *immediately following* a type?" Well, it is true that any hand played from the remaining deck should be equally likely to produce the predicted results. However, there are other things to consider. First, you would get conflicts between what you would predict based on the last hand and say, the hand before that. Priorities, and all sorts of complicated rules would have to be set up, defeating the whole concept of a simple system. Second, with only the following hand, you always make use of *the freshest* knowledge. To me this makes the most sense.

Multiple Players

Another valid question might concern the effect of more than one player on either the Min-Max or Complete System. The answer to this is:

Other players at the table will have little effect on the Min-Max or Complete System validity so long as your next hand is dealt before the deck is shuffled.

To elaborate, what difference does it make if your next hand comes from the next few cards in the deck, or from a little deeper into it after some other player gets his hand? You got your cards from the same remaining deck after your last hand, did you not? All the system says is that, *on the average,* the *remaining* deck will be favorable after certain types of hands have been played.

If the deck is to be shuffled before you bet your next hand, do not raise your bet no matter what your last hand was. In actual casino play, there will be many times when you will have to watch the dealer to see if a shuffle is imminent before you put down your bet. I found myself holding several chips and studying the dealer's hands whenever a bet-raising situation arose. If he dealt from the existing deck to the first player, I would put down the appropriately increased bet. If he shuffled, I would bet the table minimum. You will find this technique a necessary evil, especially when the table is crowded.

Also, you may feel almost foolish basing your bet increase on the results of *your* last hand when there are five other players at the table who all got different hands at the same time. Nevertheless, if you are consistent in sticking with the system, the other players' cards can actually be ignored and the system will still work. It is just as if the other players' hands were dealt to you previously while playing alone with the dealer. You ignore your own previous hands, so ignore the other players' hands as well.

What you are doing is consistently making use of the bit of knowledge gained from your own hand to size your bet. So long as you are dealt from the remaining deck, without a shuffle, you are okay and the Min-Max and Complete Systems will work. In other words, the knowledge from your hand is valuable despite the hands of other players.

Multi-Decks

You might also be concerned about the effect of multi-decks on our systems. First of all, it is not necessary to ever play in a multi-deck game anywhere in Nevada. Plenty of single-deck games are available in Las Vegas, Reno, Carson City and Lake Tahoe. They are available, but scarcer than elsewhere, on the "Strip" in Las Vegas. It is here that the multi-deck game is prominently in evidence and you may get sucked into one against both of our better judgments. Multi-decks are definitely the mode in Europe, the Bahamas, the West Indies, the Netherlands Antilles, and Puerto Rico, also.

If you do play in such a game, your overall edge will be diluted somewhat. The dilution will decrease as the remaining deck approaches fifty-two cards, on the average. That is, assuming a random distribution of high and low cards in the long run, the multi-deck game effectively reduces to a single-deck game when fifty-two remaining cards have been reached.

Another way of looking at it is this. For a two-deck game, you will start off with only half the edge you would have with one deck. As the deck is depleted, your average edge will increase linearly until you achieve the single-deck situation. Of course, these statements are valid "on the average" over many shuffles only, not every individual time the deck is shuffled.

LEANING—A BIG PROFITABLE HINT

A very simple method of greatly increasing your playing effectiveness without actually counting cards may be called "leaning." This is accomplished as follows when the table is fairly crowded.

1. **Raise your bet a little when the table is clearly littered with an excess of low cards from the last hand.**
2. **Bet the table minimum when it is very obviously covered with an excess of high cards.**

For the most part, you will not be able to tell at a glance whether there are more high or low cards on the table from the last hand. Do not worry about it, just play your system. Every so often though, everybody seems to get a lot of high cards. You do not have to count points to observe that there are pictures and aces blinding you everywhere and you see only one or two low cards. In this case, forget about your system and bet the table minimum. The deck is almost surely unfavorable.

On the other hand, if everybody is hitting and hitting like crazy, and practically nobody busts, you might see only two or three aces or pictures on the table for that go-around. Now is the time to sock-it-to-'em. Put out more than your usual maximum—maybe double it. Clearly the deck should be favorable with all these low cards gone. So "lean" in the right direction.

Since there are so many low cards on the table, the remaining deck should contain a high concentration of tens and aces. For this reason, it might be a good idea to take out insurance should the occasion arise on the next hand. This is a fine point of leaning and requires some judgment on your part.

Precisely, insurance should be taken when there are more than twice as many ten-value cards as there are non-tens remaining in the deck. Since you are not going to count and ratio tens and non-tens, you will need to exercise your intuition in this instance.

When "leaning" in this fashion, I have always exceeded my normal system expectations as predicted by the computer. It is a very natural process and is almost intuitive. Yet it is extremely effective and does not require actual counting of points. The nicest thing about it is, the more players at the table, the more effective it is. So, if all the tables are crowded, sit right down and "lean" awhile. It could be very profitable.

The Complete System is more accurate than the Min-Max method since, in reality, it represents an approximation to point counting without all that effort. Therefore, as you would expect, our initial casino results were far better than with Min-Max.

Unfortunately, the method as observed by the dealer so smacks of card counting that he very rapidly catches on to the fact that you are a threat. In fact, by the end of a single weekend my wife and I were known in all the casinos of Las Vegas. More about this in chapter 10 on casino countermeasures.

Suffice it to say that you had better move around a lot (Thorpian paper-route method) if you expect to win any money in Las Vegas.

Anyway, here is how we fared.

THE COMPLETE SYSTEM IN THE CASINO

Las Vegas

Phyll and I arrived in Las Vegas for a three-day weekend. It was our express purpose to test out a version of the Complete System while having a ball for ourselves on the town. Both objectives were accomplished to our satisfaction.

For this test, we used a modification of the complete system which gives the player about a 0.4 per cent advantage and is a lot of fun. It goes like this.

1. Bet five dollars after all pushes, splits and hard double-downs.
2. Bet one dollar after *all* four-card hands and all losses, and two-dollars after all wins, except those covered by rule one.

As you can see, this is a sort of combination of the Min-Max and Complete System. We did not bother with raising our bet after a seven-card hand or a soft double-down, since we really had not discovered these features at that time. We did not go to a five-dollar bet after split aces either, as we had correctly reasoned that this type of hand would tend to hurt the deck.

Overall, the exclusions did not hurt us seriously, and the resulting system was fun and had some virtues of its own. Importantly, the betting system is an easy one to implement since the casino chip values are typically one dollar and five dollars. For the most part, you leave your one-dollar chip and the one that the dealer pays you right in the betting box after a win. You automatically have raised your bet to two dollars for the next hand. Of course, if a five-dollar situation arises, you flip out a five-dollar chip for the next hand as you rake in anything due you. If you lose a hand, you flip out a one-dollar chip. It is a very pleasant system to play in this sense; you never have to fumble around counting chips when you are in a hurry to place your next bet. It is particularly cool in a head-to-head game with the dealer, where the action is extremely fast.

Of course, our expected edge was only about 0.4 per cent for this particular version and betting level. Had we raised our bet to ten dollars instead of five dollars after the special hands and stuck to tables with a fifty-cent minimum, we could have gotten a lot closer to an expected advantage of 1 per cent. As it turned out, we did far better than this, anyway.

We used the same estimating method as before for keeping track of our financial status and how many hands were played. Our progress is depicted below.

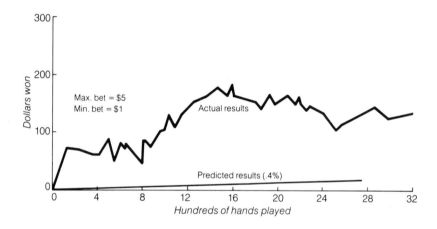

Casino Results
Complete System (Modified)

We got in about two thousand hands on that weekend and I added another five hundred hands returning from a business trip to Honolulu, and a final six hundred in Reno, Lake Tahoe and Carson City on my way home from a later trip. Again, the results were outstandingly favorable!

I actually won seventy-one dollars in my first hour of play. Do not ask me why, but once again the first hour was the best and we were never into our own money thereafter. Phyllis' records show that she broke exactly even for 110 hands during her first hour of play.

Almost every recorded data point is plotted, so you can see our best winning sessions and our worst losses on the graph. To get an idea of the time span between data points, figure about an hour of play for each 100 hands.

Both Phyll and I showed an overall profit for the weekend. Her best individual winning session yielded a fifty dollar profit over some 150 hands. Her worst sitting showed a twenty-nine dollar loss in about the same number of hands. My poorest individual session yielded a $27.50 loss in about seventy hands. Typically, we never sat at a given table much over an hour.

By the time we had played one thousand five hundred hands, the town had caught on to us, despite frequent moves to other casinos. Not really being cloak-and-dagger types, we made no attempts to disguise ourselves. Anyway, we were having too much fun observing the various casino reactions to us.

Once we were certain that we were very definitely being subjected to . . . ahh . . . countermeasures, we decided to leave the downtown area and hit the "Strip." We were spotted as we came in the door in several instances. The minute we sat down in one of these places, the dealer would be switched.

You never saw such swifties in your life as we ran into during this last session on the Strip. We would play ten or fifteen hands and always lose at an alarming rate. Often, the dealer would make comments to other players to make sure we got the message. One talented card manipulator, for instance, kept talking about how some people "fight" the game and do not play like everyone else. In other words, if you win, you are "fighting" the game.

After losing the first fourteen hands in a row in another house, I complained bitterly about the probability of that happening honestly. Do you know what the dealer matter-of-factly answered? "What do you expect? We know that you are casing the deck." Of course, I found that especially exasperating since it implied that (1) if they know you can win, they will not let you, and (2) I was counting cards, which was not true. You will find more incidents of this nature described in chapter 10, including the tale of an overzealous young dealer who inadvertently brought the entire deck down about my rapidly winning ears!

Our second objective was very easily accomplished. We caught shows with Frankie Laine, Fats Domino, Frank Sinatra, Jr. and a host of lesser lights—all in the one weekend. We had great thirty-two ounce steaks on the "Top of the Mint," a place that probably offers the view with the highest light bulb content per square foot in the whole world. We had corned beef and pastrami sandwiches for breakfast at Foxy's, and scrambled eggs at two o'clock in the afternoon. We drank free casino booze everywhere, added a couple of mugs to our collection from Aku Aku, and made frequent use of the king-sized bed in our luxurious room. We even slept once in a while —we had a ball!

Reno, Lake Tahoe and Carson City

As for these delightful areas, I vastly prefer them to Las Vegas, for playing blackjack anyway. While the rules are slightly less favorable here than in Las Vegas, the seemingly lower concentration of cheating more than makes up for it. The primary rules difference for Reno, Tahoe and Carson City are: (1) you can only double down on totals of 10 or 11; (2) except for Harrah's in Reno and the Sahara in Tahoe, insurance betting was not allowed in any of the numerous clubs in which I played.

Now, these differences cost you very little (about 0.2 per cent), and the game is a lot more pleasant, in my personal opinion, with the female dealers. For the most part, they just do not seem to care how you do; and you will rarely be cheated by them. Oh, I think most of them can get themselves a blackjack or three-card 21 after a shuffle if they get the urge, but they will not resort to this too often.

I cannot help feeling that this little trick may be the reason why insurance is disappearing in these areas. If you got wise to a first-hand blackjack-getter, for example, and that was her only defense against you, you could counteract this by taking insurance in this situation. I have occasionally done this with some success in Las Vegas, when I was sure of what was going on. Of course, if they show the 10-value card first, even the insurance possibility is washed out.

There was one club in the Lake Tahoe area, the Golden Nugget, that would let you double-down on a two-card 9. They also featured a fifty-cent minimum and fifty-dollar maximum at most tables, as well as the usual women dealers. Conditions here were quite good for the small-betting but knowledgeable noncounter. You might give it a try, especially if you are an early reader of this book.

NOTES

1. In the light of chapter 8's "General Min-Max" method, this system might be more appropriately termed "special Min-Max."

2. Wilson, *The Casino Gambler's Guide* (New York: Harper & Row, 1965), p. 156.

6. BET SIZING, BANKROLL REQUIREMENTS, AND FLUCTUATIONS IN CAPITAL

In a favorable game, the gambler's probability of ultimate ruin decreases exponentially, as his starting capital increases.[1]

FRANK SPITZER, 1964

Now that you are equipped with a simple winning system, you must understand a few other things before you even consider implementing it in a casino.

First, you will have to decide how much you can afford to *lose*. Why? Because you can never completely eliminate the possibility that this will happen. That is right, never! Oh, you can get the so-called ruin probability down exceedingly small, by choosing the proper ratio of bankroll to maximum bet size, but no matter how cautious you are, you can nevertheless be wiped out. Remember, if 100 people each play a game where they have a *90 per cent* chance of success, ten of them will lose, on the average.

Second, you should get an idea of the type of fluctuations your capital will exhibit even if you do ultimately succeed in quitting winners. I cannot really express strongly enough my own personal shock at what I have learned about this subject of fluctuations; and I do mean shock! The realities here absolutely defy your natural intuition. Many eye-opening blackjack-hand series are presented in this book, both in this chapter and the next. We will take a look at them a little later, but here is a thought for you to chew on in the meantime. As a favorable "million-hand plus" computer run shows, you would have lost in six out of 21 fifty-thousand-hand samples. This, despite the fact that the overall player's edge for the particular system tested turned out to be +0.34 per cent! Think about it. It would not be at all uncommon for you to get a fifty-thousand-hand losing streak despite the fact that you have a winning system!

BET SIZING

You will not find very much practical information published elsewhere on this subject, at least not in a form that you care to read anyway. One very big exception to this statement may be found in chapter 17 of Wilson's book, which I wholeheartedly recommend for those interested in the general subject of bank-to-bet ratio.[2] What I will present here, however, will be in keeping with the rest of this book. That is, it will be very simple and very practical information. No high altitude flights into the esoteric realms of statistics and probability will be necessary.

First, let us take a look at what kind of winning percentages and profit rates are available to us with my Min-Max and Complete System approaches. Then we will enable you to determine how much money you should take along to reduce your chances of getting wiped out to a satisfactory value. As you will see, your bankroll size and willingness to gamble will be the limiting factors for both achievable percentage and profit rate.

Percentage Edge

The most important single item to be determined is *your maximum bet size*. This parameter by itself will place an upper limit on your expected winning percentage and profit rate. Likewise, you can use your maximum bet size together with your desired "ruin probability" to determine what your bankroll must be.

To see a very vivid picture of how your expected percentage will vary with maximum bet size, refer to the following graph. In this figure it has been assumed that the table minimum is one dollar and that you will employ a maximum bet value as shown on the horizontal axis. The effect of changes in the allowable table minimum is discussed later.

For both the Min-Max and the Complete System you will notice that:

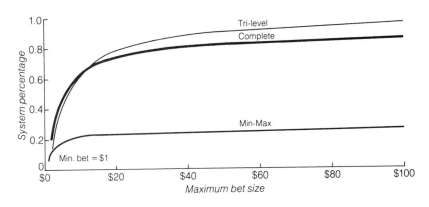

System Comparisons

Your percentage edge will increase if your maximum bet size increases.

This in itself is really not surprising, since both systems are based on gaining an advantage by increasing the bet after certain types of hands. You have merely the basic strategy edge of 0.1 per cent when flat betting, for example.

A second fact, and one that is perhaps not so obvious is that:

The maximum system percentage is approached very rapidly as your maximum bet size is increased.

Therefore, you can achieve most of the *edge* that either system has to offer without resorting to completely ridiculous maximum bets. A maximum to minimum bet ratio of 20 to 1, for instance, will get practically all of it. Even a 5 to 1 ratio will produce about half of your full potential. This could be accomplished on any fifty-cent minimum table (there are plenty in downtown Las Vegas) for a maximum bet of $2.50.

You will notice a third system plotted on the graph. This line represents what I have labeled a "trilevel" betting system which is of interest because it can raise your *percentage edge* still further. However, as we shall see, it will cut down on your actual *profit rate* somewhat. The system is based upon the fact that the computer results, as well as our own rationale, dictate that there will be *more* player's advantage after splits and doubles than after pushes and seven-card hands, although they all contribute positively. As you will recall, the pertinent statistics from chapter 4 were:

Type of Hand	Player's Edge on Following Hand
Hard Double-Downs	1.2 per cent
Split Pairs	1.4 per cent
Pushes	0.5 per cent
Seven-Card Hands	1.0 per cent

Notice that the pushes and seven-card hands yield less edge than the others. This suggests a system whereby you bet your maximum after splits and doubles and say, half your maximum after pushes and seven-card hands. This, together with betting the table minimum in all other situations, constitutes the trilevel system.

From the graph it is apparent that the trilevel system gives a slightly better percentage for maximum bets of ten dollars or more.

Below the ten-dollar level, the Complete System gives the better percentage. Frankly, since the gain is so small, and since your profit rate is significantly reduced, I do not recommend bothering with the trilevel system. It also has the added disadvantage of requiring you to fool around with a third bet size. I merely point it out because it represents about the best practical percentage you can get without counting cards.

Approximate empirical formulas for calculating your percentage edge (%E) are as follows:

System	Formula
Complete	$\%E = \dfrac{X - 0.9\,M}{1.1X + 3.2M}$
Min-Max	$\%E = \dfrac{X - 0.5\,M}{3.9X + 5.7M}$
Trilevel	$\%E = \dfrac{X - 1.1M}{X + 4.1M}$

X = Maximum bet size
M = Table minimum

Profit Rate

The mean rate at which you can expect to win money is a simple straight line function of your maximum bet size. In the graph below, the profit rates for the Min-Max, Trilevel, and Complete systems are presented. In all cases you will observe that:

Your mean profit rate will increase with a larger maximum bet size.

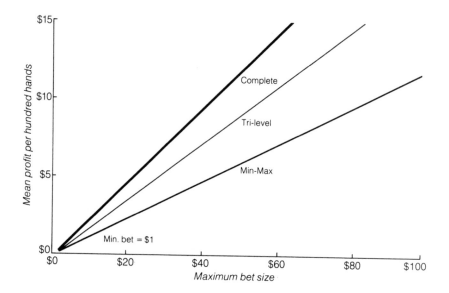

System Comparisons

The lines plotted in the graph actually represent what you can expect to win, on the average, for each one hundred hands that you play. Since you can typically play 100 hands in about an hour in the casino, these lines are also a rough estimate of what you can make per hour.

Comparing the first two graphs gives us another interesting fact. Although the trilevel system generally produces the better percentage edge for a given maximum bet size, *the Complete System always produces the maximum profit rate.* Moreover, while the percentage edge difference is uniformly small (about 0.1 per cent), the difference in profit rate can become significant. At the typical house maximum bet size of $500, for a ridiculous example, the Complete System actually provides a profit rate which is some $28.00 per hour higher than the tri-level hourly rate of about $92.00. Needless to say, if you could get a completely honest game, and play for a long time, you could really make yourself a bundle, and all without counting cards! Before you get all excited and rush off to the casino with your life's savings, however, you had better carefully consider chapter 10.

Empirical formulas for computing your mean profit per hundred hands (P_{100}) are given in the table below for each system:

System	Profit/100 Hands (Dollars)
Complete	$P_{100} = 0.24X - 0.21M$
Min-Max	$P_{100} = 0.12X - 0.05M$
Trilevel	$P_{100} = 0.18X - 0.21M$

X = Maximum bet size
M = Table minimum

Do not try to make anything meaningful out of these formulas for a maximum bet of less than two dollars.

The Effect of Changes in the Table Minimums

Since the ratio of maximum to minimum bet size is the determining factor on your available percentage edge, the house's table minimums will have a significant bearing on your chances. Now, most of the casinos in which I have played offer blackjack tables with minimum allowable bets of one dollar and five dollars. A fifty-cent minimum is fairly common in downtown Las Vegas and Reno, especially during the week. I have also seen twenty-five-dollar and twenty-five-cent minimums, but these are both relatively rare.

For a given maximum bet size, your betting ratio will necessarily be the best for the lowest available table minimum. This situation will consequently result in your most favorable system percentages and profit rates. In the next graph you will find a striking illustration of this effect for both the Min-Max and Complete systems.

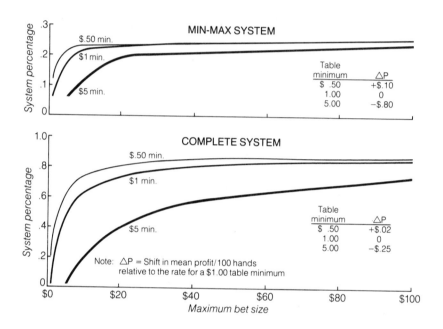

Effect of Table Minimum on System Percentage and Profit Rate

Examine the fifty-cent minimum curve and compare it to the one-dollar minimum curve just beneath it. See how much better off you are with the fifty-cent minimum at lower maximum bet levels. Further, look at how the five-dollar minimum kills you at all levels! The following point may then be emphasized.

Always play at the table with the lowest available minimum bet.

As an example of what this minimum bet can mean to your percentage edge, scrutinize the Complete System portion of the figure at the two-dollar maximum bet level. Using a half-dollar instead of a one-dollar minimum jumps your edge from 0.2 up to 0.4 per cent! Of course, if you are using a $100.00 maximum bet, it will make little difference which table minimum you use. Even a five-dollar minimum table will only cost you about 0.1 per cent of your 0.9 per cent edge at the $100.00 betting level. We might summarize these facts by stating:

A lower table minimum will significantly raise your percentage edge for a small maximum bet size.

The lower minimum will have less, but still favorable, effect on your edge at higher maximum bet levels.

Lowering the table minimum will also improve your *profit rate.* As it turns out, the improvement will be by a *constant* amount for all maximum bet sizes. Using the Complete System for an example, lowering the minimum from one dollar to fifty-cents will increase your mean profit per 100 hands by about ten cents, regardless of what your maximum bet level might be. In like fashion, raising the table minimum from one dollar to five dollars chops off a healthy eighty cents per hundred hands from your profit, no matter how large or small you make your maximum bet. So . . .

Lowering the table minimum will improve your profit rate by a constant value for all maximum bet levels.

You can obtain the constant change to your mean profit rate per hundred hands (ΔP) from the little table that follows. The change is given relative to the one-dollar table minimum and can be applied as a constant bias to the curves in the profit rate graph.

System	Table Minimum	Profit Rate Correction (ΔP)
Complete	$ .50 $1.00 $5.00	$.10 0 −$.80
Min-Max	$ .50 $1.00 $5.00	$.02 0 −$.25

A Note to Card Counters

While this book is devoted to those who wish to win without keeping track of the cards played, much of this chapter applies to the card counter as well. Since the counter also gets his edge by raising his bet in "favorable situations," he is much in the same boat as the knowledgeable noncounter.

He too can increase his profit yield by seeking out the lower table minimums. By betting fifty cents in all "unfavorable situations" he will certainly do better than he would at a five-dollar minimum table, for example. Our other remarks pertaining to optimizing percentages by virtue of higher bet values are also applicable to such a player. Of course, his profit rate will also increase linearly with increasing bet sizes in favorable situations.

DETERMINING YOUR BANKROLL REQUIREMENTS

Once you have set aside a certain amount of cash with which you can afford to gamble, then you can use this section to obtain your personal maximum bet size. Conversely, if you have your greedy heart set on a certain maximum bet size and associated profit rate, and your cash is nevertheless limited, you can find out just what your chances of going broke will be. Should the probability of losing all strike you as too large for comfort, then three courses are open to you:

1. Reduce your maximum bet to a more comfortable level.
2. Increase your playing stake.
3. Do not play at all.

The wherewithal to intelligently determine the required adjustment in the first two alternatives is provided in the subsequent discussions. Should you be weak of heart, and weaker still in the wallet, then you might seriously consider the third path, at least for the present. If you are already leaning forty-five degrees in this direction, reading chapter 10 should topple you completely off the fence.

The classic "gambler's ruin" problem and its extensive treatment by mathematicians[3] provide us with all the ammunition we need to size our stake. This information in essence involves formulas (see Appendix A) that relate your probability of ruin to your percentage edge, stake size and desired winnings. While you can calculate exact numbers for any given combination of these parameters, you should be able to quickly obtain everything that you need along these lines from the graphs and discussions that follow.

Chance of Eternal Success or Ultimate Ruin

One vital question that will concern you is this. Given a known player's edge over the house and a fixed starting capital, what are your *chances of playing forever versus ultimate loss* of your entire stake? The answer to this question is summarized in the next graph for player's edge values of 0.1, 0.25, 0.5, 1.0, 2.0 and 5.0 per cent. Obviously, the bigger your edge and the larger your starting capital, the greater are your chances of playing on eternally without ever being wiped out.

For example, if you have about a 1 per cent edge (such as with the Complete System and a maximum to minimum bet ratio greater than 10), the curve labeled 1 per cent shows that you need approximately 120 units of capital to ensure a 90 per cent chance of eternal success.

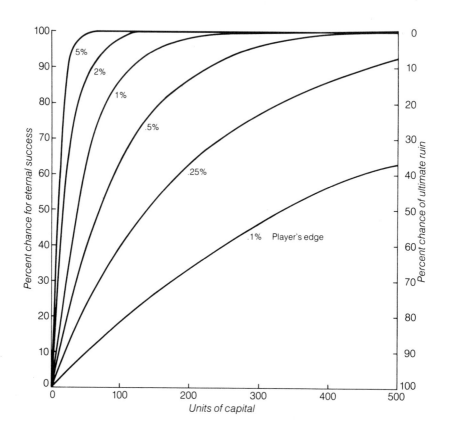

**Chance of Eternal Success
or Ultimate Ruin**

With the Min-Max System, on the other hand, you can refer to the curve labeled 0.25 per cent for an approximation. In this case, you would need approximately 460 units of capital to have the same 90 per cent chance of playing successfully for the rest of your life.

Now those curves, as are the remaining figures in this chapter, were generated from formulas that inherently assume a *flat betting* system that yields the postulated player's edge. Since our systems employ two bet levels, maximum and minimum, it is a little tricky to define what we mean by a "unit of capital." An ultraconservative approach would be to take the maximum bet as basic unit. For more realistic estimates, we should employ the average bet size. Simple formulas can be used that have been derived empirically from my computer results. These are presented below.

System	Basic Unit of Capital
Min-Max	$U = 1/5\,(2X + 3M)$
Complete	$U = 1/4\,(X + 3M)$

U = Average bet size
M = Minimum bet size
X = Maximum bet size

Notice that these formulas give precisely the correct answer for a flat bettor. Your basic unit of capital is simply your flat bet itself, for either system. As X grows large compared to M, it tends to dominate in the formulas and the basic unit of capital approaches X. The approach takes place more rapidly with the Min-Max System since more frequent maximum bets are made with it than are made with the Complete System.

For your further convenience, these formulas are plotted in the following graph for maximum bet values up to \$100 and table minimums of one and five dollars. You can use the dollar minimum line for the fifty-cent minimum case, as well, since there is little actual difference between the two in the basic unit of capital.

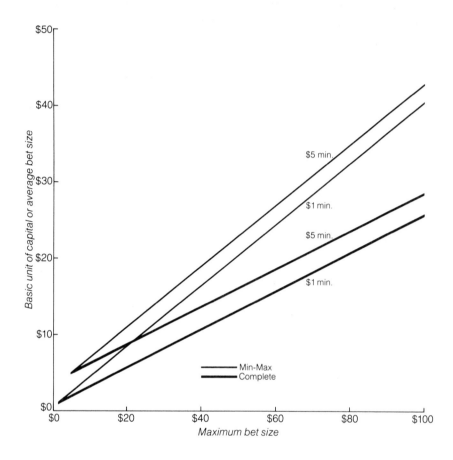

**Basic Unit of Capital
or Average Bet Size**

It is evident that the unit of capital, or average bet size, for a given maximum bet size is considerably smaller for the Complete System than it is for the Min-Max System. This means you have more units available with the Complete System for a given bankroll and maximum bet size. Consequently, your chance of ruin is correspondingly less, which is what we would expect anyway, since the Complete System is the superior one.

Chance of Achieving a Desired Profit

Another problem that is of great interest to the short-term gambler concerns the ability to achieve some finite "desired profit," given a fixed starting bankroll. Your chance of succeeding in this type of venture is naturally a strong function of your basic advantage over the house (your system percentage edge). Once again, the mathematicians' "gambler's ruin problem" considerations have produced the needed relationships (see Appendix A). These equations will enable you to calculate your exact chance of success or ruin when attempting to increase your capital a specified amount as a simple function of initial bankroll and system percentage.

If you are going to employ my Min-Max or Complete System, however, it will not really be necessary for you to perform any calculations of this nature. To spare you this effort, I have prepared the following four graphs. These graphs, together with the previous one, enable you to estimate your chances, or size your stake for systems with a player's edge of 0.1 per cent (basic strategy), 0.25 per cent (Min-Max), 0.5 per cent (intermediate bet level for Complete System) and 1 per cent (optimum Complete System with some leaning). If you are a card counter and figure you have a still higher edge over the house, then refer to the formulas given in Appendix A. It is a fairly simple matter, if you like looking up values in logarithmic tables or using scientific calculators, to calculate your own curves.

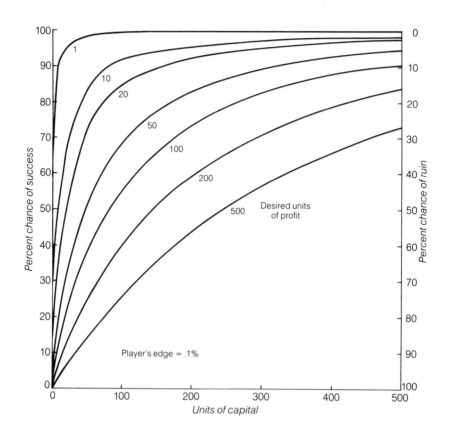

**Chance of Success
vs. Desired Profit
and Bankroll**

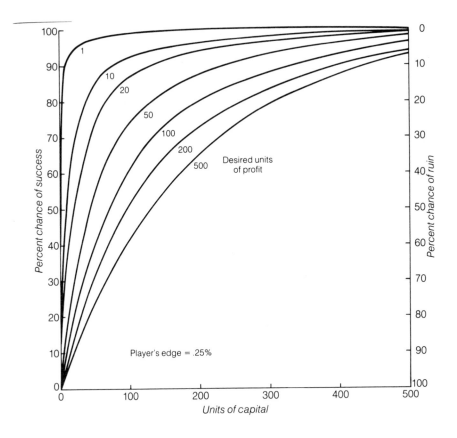

**Chance of Success
vs. Desired Profit
and Bankroll**

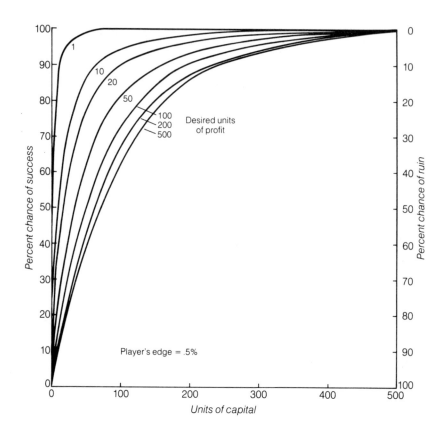

**Chance of Success
vs. Desired Profit
and Bankroll**

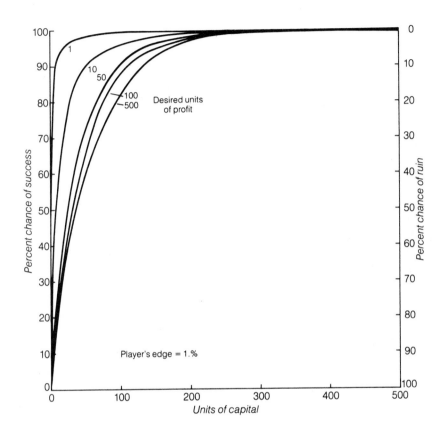

**Chance of Success vs.
Desired Profit and Bankroll**

Refer to the last of the four graphs and you will see how easy it is to make use of them. Each line, or curve, on the chart represents a "desired profit." The innermost curve corresponds to the case where you are especially greedy and wish to accumulate 500 units of profit. (A "unit of profit" is equivalent to a "unit of capital" in the previous sense). So, for this ambitious case, you can see that even with a 1 per cent edge you must have substantial capital if you wish a high probability of success. A 90 per cent chance of accumulating a 500-unit profit would again require some 120 units of capital. A desired gain of 500 units, then, is almost as ambitious as wishing to play eternally.

Proceeding outward on the graph to curves of smaller desired profit, we see that your chances of success improve considerably for a fixed starting bank. For the same 120-unit bankroll, for example, you have approximately a 95 per cent chance of accumulating a fifty-unit profit and a healthy 98 per cent chance of getting away with ten units of the casino's money. Your potential of getting ahead at least one unit is very great indeed!

Moving back to the first three graphs which are drawn for systems with a smaller player's edge, you see the same overall trends. Naturally, as your system edge is less, so then do you require more capital to go after a desired profit with the same degree of certainty that you had with the better system. This effect is especially pronounced at the greedy end. That is, the smaller system edge really kills your chance of success when seeking a large profit with a limited bankroll.

One final example should suffice to demonstrate this. Whereas 120 units were sufficient for a 90 per cent chance of success in accumulating a 500-unit profit with a 1 per cent system, you need well over 400 units to accomplish the same thing with only a 0.25 per cent edge! With the latter advantage (0.25 per cent), your 120-unit bankroll would now give you less than a 50 per cent chance of successfully acquiring a 500-unit gain.

So, now you see the utility of this chapter. From a practical standpoint, you can instantly determine just about anything you need to know about your chance of success or ruin and how it is related to your starting bankroll, desired profit, maximum bet size, and system percentage by simply reading some graphs. What more can you ask?

If you keep your percentage as high as possible and your chance of success well over 90 per cent, you can feel reasonably comfortable when you sit down to play. Nevertheless, you may be among the unfortunate few who are in for some terrible tossing on the seas of chance. Some of you, despite our best efforts, will experience some of the most demoralizing, outrageous and horrifying downward plunges of your capital imaginable. Yet, for the most part, they will be natural fluctuations that arise without any help from an artful dealer. If you are the one in a hundred that sat down with a 99 per cent chance of success and got skinned anyway, do not say I did not warn you. It can and will happen! Read on to gain a little appreciation of the possible effect of natural statistical fluctuations on your carefully laid plans.

FLUCTUATIONS IN CAPITAL

My subsequent discussion of fluctuations will probably make the statisticians among you scream in agony. I do not care. For the rest of you, there is some graphic, meaningful information that you should enjoy absorbing. I will not crush you with mean values, standard deviations and various statistical tests of the data.

Not that this type of information is not useful to scientists. It is. I have used just this sort of statistical information, and much more, in my work in orbit determination and satellite tracking, for example. The point is, we wish to avoid losing your interest with a boring presentation of what should be fascinating material.

A Look at the Long Run

Let me ask you a question. How many hands of blackjack would you consider to be a large, meaningful sample? One thousand? Ten thousand? Fifty thousand? Well, that is a tough question and a loaded one. As you will see, even a fifty-thousand-hand sample will often give you some misleading statistics.

My own personal definition of the "long run" is that quantity of trials that dependably produces meaningful statistics. Let me tell you, I have analyzed a lot of blackjack hands. This includes millions on the computer, tens of thousands manually dealt for laboratory experimentation and many thousands in the casinos. Out of all of this experience has come a definite awe for the wildness of natural statistical fluctuations. So, if you wish to determine any statistics on a particular system, if you desire to get in the "long run" with it, you must look at a great many hands indeed.

For an example, I have chosen a fairly well-behaved computer sample of one million hands. In the next graph you will find the profit history for two systems. The upper trace is for a modified version of the complete system that gives the player about a 0.34 per cent edge. This graph is for the same technique that was used in our Nevada tests of the Complete System (see chapter 5). One data point was plotted for every fifty thousand hands the computer played. While the overall system percentage (+0.34 per cent) is clearly validated by this run, the graph also harbors a horrifying corollary. Six of these twenty samples of fifty thousand hands each were losing streaks!

Fifty-thousand-hand losing streaks are common, even when you have an edge as large as +0.34 per cent.

As a matter of fact, this particular run shows a one-hundred-fifty-thousand-hand losing streak starting at eight hundred thousand hands. And this is nice, well-behaved data. I have seen much, much worse than this—and all perfectly valid. Yes, you can have the edge, but unless you intend to play an awful lot of hands, you can certainly lose anyway.

Also plotted in the graph are the basic strategy results for a flat one-dollar bet. This run gave very gratifying results, never straying far from the expected +0.1 per cent player's edge. Again, I have other runs which contain basic sequences which I am still unable to completely accept despite certain knowledge of their validity.

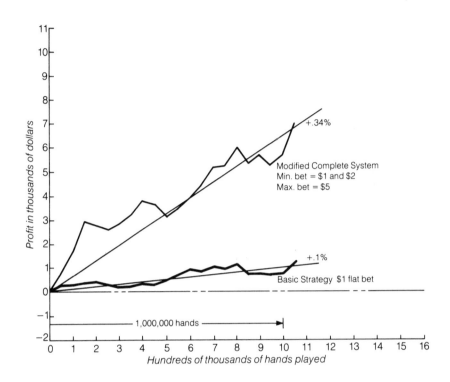

The Long Run
One Million Hands of
Blackjack

I am by no means alone in the scientific community with a healthy respect for fluctuations. The basic strategy is an essentially even game, as you know, and therefore is roughly equivalent to a coin-tossing problem. Feller has this to say on that subject:

The results concerning fluctuations in coin tossing show that widely held beliefs about the law of large numbers are fallacious. These results are so at variance with common intuition that even sophisticated colleagues doubted that coins actually misbehave as theory predicts.[4]

If you think that playing the basic strategy guarantees that you will sooner or later gain the lead over the house, you still have not gotten the message. If you start off losing, you may go on losing forever. But, as Feller once again tells us, you disbelievers have a lot of company:

Few people will believe that a perfect coin will produce preposterous sequences in which no change of lead occurs for millions of trials in succession, and yet this is what a good coin will do rather regularly.[5]

If all of this seems like a monument to negativity to you, it is for a very good reason. Despite the fact that you now have a beautiful winning system in your possession, you may still get fluctuated out of your money in the casinos if you are not careful. Use this chapter meticulously and keep that ruin probability way down in the noise.

Short-Term Fluctuations

Since you will probably not play one million, or even fifty thousand hands, a closer look at the "short run" is perhaps more pertinent to your cause. To this end, take a good look at the graph of short-term fluctuations. Here are plotted the results for twenty thousand hands of blackjack at a rate of one point every 100 hands. Thus you can get a feel for how your status will vary in about one-hour increments.

The dotted curve is a profit history for the +0.34 per cent system previously described. The dashed curve is for the one-dollar flat bet basic strategy. The first ten thousand hands are shown in the lower half of the figure and the results between ten and twenty thousand hands are depicted on the upper half.

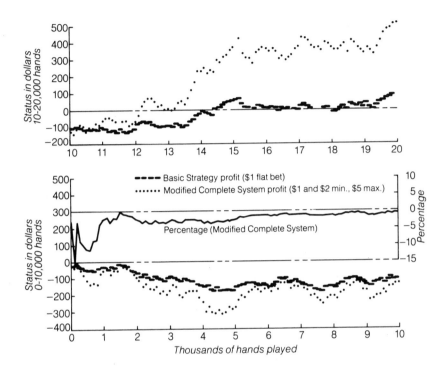

Short Term Fluctuations
A Blow-up of 20,000 Hands
of Blackjack

This is a particularly good sample for demonstration purposes as it shows some normal positive and negative fluctuations. As you can see, a substantial losing session took place for the first forty-five hundred hands (roughly a forty-five-hour losing streak). The overall basic strategy edge for the twenty thousand hands was +0.39 per cent and the modified complete system produced +1.27 per cent in this sample. So you see, this is a very favorable set of data for the player, and yet you would have been in the red for some twelve thousand hands with the better system and for 14,500 hands with the basic strategy alone!

This sample, incidentally, was not chosen for its fluctuations, its final results, or anything of that nature. Rather, it was just the first twenty thousand hands spewed out by the computer on the million-hand run used for the previous graph. It is simply typical. Much wilder sequences abound in the data.

As you can see, no particular one-hour session will affect you drastically at these low betting levels. Changes in either direction in excess of forty or fifty dollars are very rare for both systems. But when you start putting these one-hour sessions together, the collective result can be quite shattering to the bettor.

Also shown is the cumulative system percentage for the first ten thousand hands of the modified Complete System. Using the scale on the extreme right, you will notice that the first 100 hands of play produced a system percentage of –15 per cent. A large fluctuation? Not really, as this is very typical for such a small sample. As the hands build up and the sample size grows, however, you can see that the fluctuations in cumulative system percentage smooth out. By the time ten thousand hands have gone by, the percentage has already converged to within about 1 per cent of the true value. What this points out is an important fact about the law of large numbers:

As the sample size (number of hands) gets large, the true system percentage will be approached.

Not so obvious, however, is this concept:

While the true system percentage is consistently approached more closely with growing sample size, the actual profit may continue to diverge from the expected value.

If this seems contradictory to you, refer once again to the graph. Notice that in the first 4,500 hands, the modified complete system profit went from zero to –$295.00. At the same time that this loss was taking place, the percentage improved from –15 per cent at 100 hands to –3 per cent at 4,500 hands. This is because the total investment grew more rapidly than the change in profit and *that* then, is the key to the whole thing. While your profit fluctuates, your total investment builds up steadily. Profit fluctuations have less and less effect on percentage as the sample size and total investment increases, and the true percentage is ultimately approached.

So, now you have the important facts. Use this chapter well and perhaps you can survive the terrifying statistical violence to which you will certainly be subjected. Let us hope that things fluctuate your way.

NOTES

1. Frank Spitzer, *Principles of Random Walk* (New York: D. Van Nostrand and Co. Inc., 1964), p. 218.

2. Wilson, *The Casino Gambler's Guide* (New York: Harper & Row, 1965), pp. 259–273.

3. You can find this problem discussed in the following references, to name a few: Feller, Wilson, Loeve, Parzen, Spitzer. See bibliography.

4. William Feller, *An Introduction to Probability Theory and Its Applications*, Vol. I (2nd ed.; New York: John Wiley & Sons, 1957), p. viii.

5. Ibid., p. 81.

7. DOING YOUR OWN BLACKJACK RESEARCH BY HAND

I debated at length with myself about including this chapter at all. Finally, thinking back on how much fun we had playing off our twenty thousand hands and analyzing the results, and realizing that there may be many of you who are laid up or hospitalized and would love to fool around with blackjack in this fashion, I decided to include it.

Nowhere in the literature can you find this type of thing for studying blackjack. Since we had to develop these techniques from scratch, perhaps it can be of benefit to those of you who are interested in developing and testing some new blackjack system without benefit of a computer. Those of you who are not so disposed may ignore this chapter.

PICKING A PLAYING STRATEGY

Unless you are going to employ card counting, I would definitely recommend the basic strategy of chapter 3. Keeping your Du-Rite

Wheel handy is an excellent way of insuring that you make no playing errors.

RECORDING THE DATA

It is important that you go about your research systematically and with extreme care. Otherwise, you may find yourself loaded down with a bushel basket full of useless papers, notes and generally disorganized data. To insure that your data are accurate, I definitely recommend that you play your test hands with another person. This helpmate can insure that your theoretical reactions, as well as those of the hypothetical dealer, are in strict agreement with the club rules and your specified playing techniques.

Wins, Losses and Pushes

A good technique, I found, was to have one person lay out the cards, while the other one records each hand as either a win (W), or loss (L), or push (P) for the player. For example, I laid out both my hand and the dealer's while my wife recorded the Ws, Ls and Ps.

A sample page is reproduced here. The sequence of hands has been numbered from one to fifty-one so that we may refer to particular hands of interest with ease. You will *not* need to number your hands in this manner.

Runs

To enable an accurate and easy computation of the profit achieved by a given set of hands, it will be best if you divide your data into runs. This will allow you to use the profit equations presented a little later. That is, mark off the divisions between groups of winning hands and losing hands. Ignore those hands labeled as pushes (Ps) when identifying runs. For example, hands one through ten on our sample page would be grouped as follows:

CUMULATIVE STATISTICS
Won = 2623
Lost = 2822
Total hands = 5445

No.	No.	No.	No.
1. W a_1	15. W	29. W	43. W
2. P	16. W	30. W* (Y_1B)	44. WD (Y_2WD)
3. L	17. W	31. W a_3	45. P
4. LD (X_1LD)	18. W a_6	32. L b_1	46. W a_4
5. L b_3	19. L b_1	33. W	47. L b_1
6. W	20. W a_1	34. W a_2	48. W
7. S$_W^W$ (Y_1WS)	21. L b_1	35. L	49. W a_2
8. W a_3	22. W	36. S$_{LD}^L$ (X_1LS) (X_1LD)	50. L b_1
9. L b_1	23. W	37. L b_3	51. W a_1
10. W a_1	24. W a_3	38. W a_1	
11. L b_1	25. L b_1	39. L	
12. P	26. W* (X_1B)	40. L	
13. W	27. W a_2	41. L b_3	
14. W* (Y_1B)	28. L b_1	42. W	

STATISTICS FOR THIS PAGE			
a_1 = 5	b_1 = 9	X_1B = 1	Y_1B = 2
a_2 = 3	b_2 = 0	X_1LD = 2	Y_1WS = 1
a_3 = 3	b_3 = 3	X_1LS = 1	Y_2WD = 1
a_4 = 1			
a_5 = 0			
a_6 = 1			

Sample Record With Statistics

$\overline{\underline{W}}$ (one win)
P
L
L (three losses in a row)
$\underline{L}$
$\overline{W}$
W (three wins in a row)
W
$\underline{L}$ (one loss)
$\underline{W}$ (one win)

Doubled Hands

If, on a particular hand, we went double down, we recorded that fact by placing a (D) after the record of the hand's outcome. Thus, a situation where you doubled and won would be entered as WD (win double). If you doubled and lost, it would be entered as an LD (lose double). For example, referring to our sample page, hand number four was a case where I doubled and lost. The first five hands of the page would now read:

$\underline{W}$
P
L
LD
$\underline{L}$

The only other doubling hand on this page was hand number forty-four which was a win. This hand was recorded as a WD (win double).

Split Pairs

We found that it was clearest to record split pairs as a single (S). The outcome of each portion was then marked alongside the S. For two wins, for example, this would read (S_W^W), for two losses (S_L^L), and for a win and loss (S_L^W).

Doubled hands originating from a split pair may be indicated, as before, with a capital D. Thus a split pair ending up in a loss and a doubled win would read (S^L_{WD}). This notation may seem a bit awkward but it is necessary for purposes of accuracy.

Note that a problem exists here with respect to dividing your data into runs. Which hand shall you consider in the event of a split resulting in one loss and one win? We solved this dilemma by always taking the win for the division purposes and treating the loss as an isolated bet. Of course, no problem exists for two wins or two losses occurring as the result of a split.

On our sample page, hand number seven was a "win split" (two wins) and recorded as (S^W_W). Similarly, hand thirty-six was a "lose split," one hand being doubled. This was entered as (S^L_{LD}), in this case, the LD was treated as the isolated bet.

Blackjack or Naturals

In those cases where the player is dealt an ace and a ten or an ace and a picture card, the hand is recorded as a win (W). To identify it as a natural, an asterisk (*) may be added so that the entry appears as (W*) and is read "win blackjack." Three blackjacks occurred on our sample page. These were on hands fourteen, twenty-six and thirty.

COMPUTING THE PROFIT

Labeling the Runs

In order to use the profit equations to compute the profit from your set of wins and losses, you must know how many of each kind of run occurred. That is, how many one in a row, two in a row, et cetera, runs of wins and similar runs of losses occurred. Therefore, you should adopt the following notation:

TOTAL NO. OF WINS IN A ROW	SYMBOL	TOTAL NO. OF LOSSES IN A ROW	SYMBOL
1	a_1	1	b_1
2	a_2	2	b_2
3	a_3	3	b_3
4	a_4	4	b_4
5	a_5	5	b_5
6	a_6	6	b_6
7	a_7	7	b_7
.	.	.	.
.	.	.	.
.	.	.	.
n	a_n	n	b_n

Thus you can call the total number of losing runs that occurred, say "three in a row" as b_3; and similarly the total number of winning runs of "ten in a row" will be denoted by a_{10}. The symbol a_i, therefore, refers to the total number of winning runs of length i and the symbol b_i refers to the total number of losing runs of length i.

Now, looking at your sample page again, you will see that alongside each group was placed the symbol a_i or b_i into which *total* the individual run will enter. You know, for example, that the first hand constitutes a win of "one in a row," and therefore it will add one to the *total* number of one in a row wins that is denoted by a_1. Likewise, hands six, seven and eight constitute a winning run of three in a row and will add "one" to the *total* denoted by a_3.

The pushes (P), or tied hands, were ignored completely when marking off and labeling the runs.

Labeling the Bets

It will also be necessary for computing the profit to assign a value for each bet made. For instance, let the bets starting with a first win of a winning run be equal in dollar value to Y_0, Y_1, Y_2, et cetera, as shown below:

Outcome	Bet ($)
L	
W	Y_0
W	Y_1
W	Y_2
W	Y_3
.	.
.	.
W	Y_n
L	

Similarly, in a losing series, let the dollar values be X_0, X_1, X_2, et cetera, as shown:

Outcome	Bet ($)
W	
L	X_0
L	X_1
L	X_2
L	X_3
.	.
.	.
.	.
L	X_n
W	

The X and Y values are the dollar values won or lost for the corresponding outcome. Thus, Y_0 dollars were won on the first win of a series, Y_1 dollars were won on the second win and so on. Similarly X_0 dollars were lost on the first loss of a series, X_1 dollars on the second loss, and so on.

It should be fairly obvious that the first quantities X_0 or Y_0 in any series is a variable that depends upon what kind of a run is being closed out. In any case, this fact is discussed in detail in Appendix B. All other X and Y values are fixed constants for all runs.

So, the value of X_1 in a run of five losses is the same as the value of X_1 in a run of seven losses, or three losses, or any other size losing run.

To further enhance your understanding, let us label the amounts won and lost on the first ten hands of our sample.

We have:

Outcome	Amount Won or Lost
W	Y_0
P	
L	$X_0 = Y_1$
L	X_1
L	X_2
W	$Y_0 = X_3$
W	Y_1
W	Y_2
L	$X_0 = Y_3$
W	$Y_0 = X_1$

The variability of the X_0 and Y_0 bets is amply demonstrated by this series of runs.

It will not be necessary to label all of your hands in this fashion. All that is necessary is that you understand what has just been covered. You will see that identifying the dollar value of the hands will be required only for those hands that fall into a category that we will call "extra money." This is covered in the next section.

Extra Money

Every time you double down you either win or lose some additional money. Likewise, when you split your pairs or win a natural, you lose or win more, or less, money than for an ordinary hand. These three types of hands may be categorized as "extra money" hands and must be accounted for in our computation of the profit.

DOUBLE DOWN. Every "win double" results in one extra win corresponding to the Y value assigned. Every "lose double" similarly results in one additional loss corresponding to the X value assigned to it. One-half of the total doubled bet will be treated as a normal W or L in our profit computation and will add to the particular run in which it occurs. The "extra" win or loss will be handled separately and its dollar value (position in the run in which it occurs) must be identified correctly. We adopted the following notation for doubled hands:

Y_1WD = Total number of "win doubles" equal in value to Y_1
Y_2WD = Total number of "win doubles" equal in value to Y_2
Y_3WD = Total number of "win doubles" equal in value to Y_3

.

Y_nWD = Total number of "win doubles" equal in value to Y_n
Y_1LD = Total number of "lose doubles" equal in value to Y_1
Y_2LD = Total number of "lose doubles" equal in value to Y_2
Y_3LD = Total number of "lose doubles" equal in value to Y_3

.

Y_nLD = Total number of "lose doubles" equal in value to Y_n
X_1WD = Total number of "win doubles" equal in value to X_1
X_2WD = Total number of "win doubles" equal in value to X_2
X_3WD = Total number of "win doubles" equal in value to X_3

.

X_nWD = Total number of "win doubles" equal in value to X_n
X_1LD = Total number of "lose doubles" equal in value to X_1
X_2LD = Total number of "lose doubles" equal in value to X_2

.

X_nLD = Total number of "lose doubles" equal in value to X_n

Actually it is quite easy to identify which total symbol (X or Y) applies to a particular doubled hand.

Look, for example, at hands forty-one to forty-seven on our sample page. We see that hand forty-four was a "win double." The hand outcomes and bet values are tabulated below:

Hand Number	Outcome	Bet Value
41	L	X_2
42	W	$Y_0 = X_3$
43	W	Y_1
44	WD(Y_2WD)	Y_2
45	P	———
46	W	Y_3
47	L	$X_0 = Y_4$

The *win double* occurred on a Y_2 bet. Therefore, it is added into the total figure denoted Y_2WD. Had the win doubles occurred on hand forty-six, it would have gone into the total denoted Y_3WD since the bet value was Y_3. It is as simple as that! You should be able to tell, after a brief practice, whether a particular win double is of the Y_2 type, the Y_5 type, or whatever, just by looking at its position in the run. Each win double or lose double adds one to its appropriate Y or X value.

You will notice that, had hand forty-two been a "win double," the quantity one would have been added to the total figure X_3WD since it came up on an X_3 bet. Again, had hand forty-one been a lose double, the quantity one would be added to the total figure X_2LD.

SPLIT PAIRS. For split pairs, precisely the same procedure was followed. The most favorable hand of the split pair is absorbed by the system and goes into the normal makeup of the runs. The other hand is treated as "extra money" exactly as with the extra hands from doubling. The notation, which is essentially the same, is:

Y_1WS = Total number of "win splits" equal in value to Y_1
Y_2WS = Total number of "win splits" equal in value to Y_2
.

Y_nWS = Total number of "win splits" equal in value to Y_n
Y_1LS = Total number of "lose splits" equal in value to Y_1
Y_2LS = Total number of "lose splits" equal in value to Y_2
.

Y_nLS = Total number of "lose splits" equal in value to Y_n
X_1WS = Total number of "win splits" equal in value to X_1
X_2WS = Total number of "win splits" equal in value to X_2
.

X_nWS = Total number of "win splits" equal in value to X_n
X_1LS = Total number of "lose splits" equal in value to X_1
X_2LS = Total number of "lose splits" equal in value to X_2
.

X_nLS = Total number of "lose splits" equal in value to X_n

BLACKJACKS. Finally, blackjacks or naturals are tallied up in similar fashion. The notation is:

Y_1B = Total number of blackjacks occurring on a Y_1
Y_2B = Total number of blackjacks occurring on a Y_2
.

Y_nB = Total number of blackjacks occurring on a Y_n
X_1B = Total number of blackjacks occurring on a X_1
X_2B = Total number of blackjacks occurring on a X_2
.

X_nB = Total number of blackjacks occurring on a X_n

Since a natural pays 3 to 2, the actual "extra money" added by a blackjack is equal in dollars to one-half of the bet made. Thus, you will see that the "extra money" terms for blackjacks in the profit computations are all divided by two.

Cumulative Statistics

At the top of the sample page you will find "Cumulative Statistics." These are:
Won = Total number of hands won, up to and including this page.
Lost = Total number of hands lost, up to and including this page.
Total hands = Won + lost.

Individual Statistics

At the bottom of the sample page you will observe a summary of the statistics (*a*s, *b*s, *X*s, and *Y*s) for that page alone. These data will be added up with similar data from all the other pages to get the final values of the statistics. They are all needed for the profit computations which are next described.

THE PROFIT EQUATIONS

In this section you will learn how to compute the profit accrued by *any betting system* for the strategy you employed in compiling your data. Now you are ready to reap the benefits of all your painstaking research!

The formulas presented in this chapter *are simple* to apply. The derivations of these equations are given in detail in Appendixes B and C of this book.

It is *not* at all necessary that you understand the derivations. Just plug the statistics that you have compiled into the formulas supplied and you can compute the profit with ease. All you need to know for this is simple arithmetic. Do not let the appearance of the equations frighten you off! Plug in the numbers! That is all you have to do.

Formulas

Given:

These quantities are obtained by totaling up the statistics from your individual ledger pages.

$$a_1, a_2, a_3, a_4, \ldots\ldots\ldots\ldots\ldots\ldots\ldots\ldots\ldots\ldots\ldots\ldots\ldots a_n$$
$$b_1, b_2, b_3, b_4, \ldots\ldots\ldots\ldots\ldots\ldots\ldots\ldots\ldots\ldots\ldots b_n$$
and:
$$X_1WD, X_2WD, \ldots\ldots\ldots\ldots\ldots\ldots\ldots\ldots\ldots\ldots X_nWD$$
$$X_1LD, X_2LD, \ldots\ldots\ldots\ldots\ldots\ldots\ldots\ldots\ldots\ldots X_nLD$$
$$X_1WS, X_2WS, \ldots\ldots\ldots\ldots\ldots\ldots\ldots\ldots\ldots\ldots X_nWS$$
$$X_1LS, X_2LS, \ldots\ldots\ldots\ldots\ldots\ldots\ldots\ldots\ldots\ldots X_nLS$$
$$X_1B, X_2B, \ldots\ldots\ldots\ldots\ldots\ldots\ldots\ldots\ldots\ldots\ldots X_nB$$
$$Y_1WD, Y_2WD, \ldots\ldots\ldots\ldots\ldots\ldots\ldots\ldots\ldots\ldots Y_nWD$$
$$Y_1LD, Y_2LD, \ldots\ldots\ldots\ldots\ldots\ldots\ldots\ldots\ldots\ldots Y_nLD$$
$$Y_1WS, Y_2WS, \ldots\ldots\ldots\ldots\ldots\ldots\ldots\ldots\ldots\ldots Y_nWS$$
$$Y_1LS, Y_2LS, \ldots\ldots\ldots\ldots\ldots\ldots\ldots\ldots\ldots\ldots Y_nLS$$
$$Y_1B, Y_2B, \ldots\ldots\ldots\ldots\ldots\ldots\ldots\ldots\ldots\ldots\ldots Y_nB$$

These may be varied to reflect any betting system desired!

$X_1, X_2, X_3,$ .... X_n

$Y_1, Y_2, Y_3,$ .... Y_n

Compute:

$$A_0 = a_1 + a_2 + a_3 + \ldots + a_n \qquad B_0 = b_1 + b_2 + b_3 + \ldots + b_n$$

$$A_1 = A_0 - 2a_1 \qquad\qquad\qquad B_1 = B_0 - 2b_1$$

$$A_2 = A_1 + a_1 - 2a_2 \qquad\qquad B_2 = B_1 + b_1 - 2b_2$$

$$A_3 = A_2 + a_2 - 2a_3 \qquad\qquad B_3 = B_2 + b_2 - 2b_3$$

$$A_4 = A_3 + a_3 - 2a_4 \qquad\qquad B_4 = B_3 + b_3 - 2b_4$$

.

.

.

$$A_n = A_{n-1} + a_{n-1} - 2a_n \qquad B_n = B_{n-1} + b_{n-1} - 2b_n$$

Set 1

$$A_{1T} = A_1 + (Y_1 WD + Y_1 WS + .5 Y_1 B - Y_1 LD - Y_1 LS)$$

$$A_{2T} = A_2 + (Y_2 WD + Y_2 WS + .5 Y_2 B - Y_2 LD - Y_2 LS)$$

$$A_{3T} = A_3 + (Y_3 WD + Y_3 WS + .5 Y_3 B - Y_3 LD - Y_3 LS)$$

$$A_{4T} = A_4 + (Y_4 WD + Y_4 WS + .5 Y_4 B - Y_4 LD - Y_4 LS)$$

.

.

.

$$A_{nT} = A_n + (Y_n WD + Y_n WS + .5 Y_n B - Y_n LD - Y_n LS)$$

$$B_{1T} + B_1 - (X_1 WD + X_1 WS + .5 X_1 B - X_1 LD - X_1 LS)$$

$$B_{2T} = B_2 - (X_2 WD + X_2 WS + .5 X_2 B - X_2 LD - X_2 LS)$$

$$B_{3T} = B_3 - (X_3 WD + X_3 WS + .5 X_3 B - X_3 LD - X_3 LS)$$

$$B_{4T} = B_4 - (X_4 WD + X_4 WS + .5 X_4 B - X_4 LD - X_4 LS)$$

.

.

.

$$B_{nT} = B_n - (X_n WD + X_n WS + .5 X_n B - X_n LD - X_n LS)$$

Set 2

$$W_T = A_{1T} Y_1 + A_{2T} Y_2 + A_{3T} Y_3 + \ldots\ldots +A_{nT} Y_n + a_0 Y_0$$

$$L_T = B_{1T} X_1 + B_{2T} X_2 + B_{3T} X_3 + \ldots\ldots +B_{nT} X_n + b_0 X_0$$

$$P_T = W_T - L_T$$

Set 3

SAMPLE PROFIT CALCULATIONS

To illustrate how the profit equations are used, let us calculate the profit for the statistics of the sample page. Assume that this page of statistics is all that we have.

Hand number one is the first one bet. Since this first hand does not close out any preceding run, it actually represents a Y_0 bet. Moreover, it provides your statistic (a_0), which is equal to one. The statistic b_0 is equal to zero. Had the first hand been a loss, then a_0 would have been zero and b_0 would have been one. These are the only allowable values for a_0 and b_0 since there can only be one first hand for each complete set of data. If, by chance, the first bet should contain "extra money," then a_0 and b_0 should be adjusted accordingly.

Copying the other statistics from the bottom of the sample page we have:

Given:
$$a_0 = 1 \quad b_0 = 0 \quad X_1B = 1 \quad Y_1B = 2$$
$$a_1 = 4 \quad b_1 = 9 \quad X_1LD = 2 \quad Y_1WS = 1$$
$$a_2 = 3 \quad b_2 = 0 \quad X_1LS = 1 \quad Y_2WD = 1$$
$$a_3 = 3 \quad b_3 = 3$$
$$a_4 = 1$$
$$a_5 = 0$$
$$a_6 = 1$$

From Set 1

$$A_0 = 12 \qquad\qquad B_0 = 12$$
$$A_1 = 12 - 8 = 4 \qquad\qquad B_1 = 12 - 18 = -6$$
$$A_2 = 4 + 4 - 6 = 2 \qquad\qquad B_2 = -6 + 9 - 0 = 3$$
$$A_3 = 2 + 3 - 6 = -1 \qquad\qquad B_3 = 3 + 0 - 6 = -3$$
$$A_4 = -1 + 3 - 2 = 0$$
$$A_5 = 0 + 1 - 0 = 1$$
$$A_6 = 1 + 0 - 2 = -1$$

From Set 2

$$A_{1T} = 4 + (1 + 1) = 6 \qquad\qquad B_1 = -6 - (0.5 - 2 - 1) = -3.5$$
$$A_{2T} = 2 + (1) = 3 \qquad\qquad B_2 = 3$$
$$A_{3T} = -1 \qquad\qquad B_3 = -3$$
$$A_{4T} = 0$$
$$A_{5T} = 1$$
$$A_{6T} = -1$$

Now, let us further assume that your betting system is such that you progress when you win and also progress when you lose. Let us start with $X_0 = Y_0 = \$2.00$. Then, for each additional win or loss in a series add $2.00 as follows:

Winning Runs		Losing Runs	
Outcome	*Bet*	*Outcome*	*Bet*
W	$Y_0 = \$2$	L	$X_0 = \$2$
W	$Y_1 = \$4$	L	$X_1 = \$4$
W	$Y_2 = \$6$	L	$X_2 = \$6$
W	$Y_3 = \$8$	L	$X_3 = \$8$

From Set 3

$W_T = 1(2) + 6(4) + 3(6) - 1(8) + 0(10) + 1(12) - 1(14) = \$34.00$
$L_T = 0(2) - 3.5(4) + 3(6) - 3(8) = -\$20.00$
$P_T = \$34.00 - (-\$20.00) = \$54.00$
$$P_T = \$54.00$$

For this sample of data, then, your simple betting technique would have won $54. To verify this yourself, try playing through the sample page, placing your bets with poker chips. For each win take the appropriate winnings from the "bank" of chips. For each loss, place the amount of your bet in the bank. If you are careful and observe the doubles, splits and blackjacks, and pay yourself or the bank accurately, you will find that you have $54 more worth of chips than when you started.

Incidentally, I definitely do *not* recommend that you use this betting system of progressing on winning *and* losing runs. Almost any sensible method of betting would produce winnings for this particular sample since it was a very favorable set of data.

In the next chapter, we will examine many different types of betting systems. Perhaps the analyses will give you some ideas for constructing your own system. If so, you can use the methods of this chapter to build yourself a good data base.

8. OTHER SYSTEMS APPLIED TO BLACKJACK

A great variety of betting systems have been devised by gamblers in the hope of becoming a surefire winner. Unfortunately, the bulk of these systems are based upon fallacious logic and are ultimately disastrous to the user. In fact, when applied to such casino games as craps, roulette or keno, they are *all* losers.

A select few of the systems that I have unearthed in the literature do have some merit for the purpose of good money management however, when applied to basic strategy blackjack. Again, these are all noncard-counting systems that are dealt with in this chapter. Counting systems are dealt with subsequently.

Now all of you have heard or read pompous statements by scientists, mathematicians and the like to the effect that mathematical theory shows that *all* gambling systems are doomed to failure when the house has the edge. While I must concur, I cannot help but feel the same reaction that most of you undoubtedly have to this. Namely, what do you care about mathematical theory? You have a system that intrigues you and what is really of interest is testing it out on a lot of real data. Furthermore, if you play the basic strategy, the game is about even. Maybe now your favorite system will work. Right? Maybe, but if you read on you may find out differently.

This chapter, then, will accomplish two objectives.[1] First, it will provide you with a simple tool for testing out almost any non-counting betting system on twenty thousand hands of basic strategy blackjack, and in just a few minutes. Second, it will show you graphically how every system that I could find in the literature, and some variations which I concocted myself, would fare in the same twenty thousand-hand test.

Of course, a twenty thousand-hand test is not sufficient to establish the validity of a system. It is, in most cases though, quite adequate to demonstrate the kind of trouble you may get into when playing your system. This sample size will give you a good general feeling for the type of short-term fluctuations to expect and will provide a yardstick for comparing one system to another.

A TWENTY THOUSAND-HAND TEST YOU CAN TRY YOURSELF IN JUST A FEW MINUTES

Over the course of several months, Phyll and I actually played and recorded twenty thousand hands of blackjack under rigorous casino rules. I dealt and played the hands adhering to the basic strategy and casino rules. Phyll recorded the results in a ledger and carefully checked each hand for accuracy of play. These hands were all dealt to one player only, and from a single fifty-two-card deck. This exercise, while quite grueling, provided a good control for my computer experimentation and a tremendous first-hand "feel" for the basic game.

While my computer program could have played and recorded this many hands in forty-eight seconds or less, the validity of data carefully generated by hand cannot be seriously questioned by anyone. For this reason, these data have been used exclusively for providing you with a testing method and in comparing the various systems. If you desire, you can extend the data yourself using the methods described in chapter 7.

The Profit Coefficients

In the following table you will find a set of "profit coefficients" tabulated at 500-hand intervals. You can use them to very quickly determine how any "run based" system would have performed over this data sample. Now, a run is just a series of wins or losses in a row. Another applicable name for such systems would be "sequence based systems."

The numbers in the table, therefore, may be used for any system wherein you change your bet systematically after any number of wins or losses. The double-up when you lose system, or "Small Martingale," is a "run based" system in this sense. Other examples of run based systems, all of which are examined in this chapter, are the "Grand Martingale," "Goodman's Progressions," the "Oklahoma System," my own "General Min-Max" technique, "Limited Martingales" and a number of interesting variations.

TOTAL HANDS PLAYED

	500	1,000	1,500	2,000	2,500	3,000	3,500	4,000	4,500	5,000
A_1	6.0	-7.5	1.5	-12.5	-19.5	-10.0	-14.0	1.0	-2.5	-5.5
A_2	8.0	14.5	22.5	12.0	6.5	-0.5	12.5	18.0	14.5	29.5
A_3	12.5	7.5	11.5	10.5	13.5	13.5	8.0	10.0	6.0	-1.0
A_4	7.0	13.5	19.0	19.0	24.0	30.5	34.0	38.5	37.0	40.0
A_5	6.5	4.5	6.0	4.5	1.5	0.0	-1.0	-9.0	-6.0	-3.5
A_6	-2.0	0.0	2.0	2.0	2.0	1.5	3.0	4.0	8.0	10.0
A_7	1.0	0.0	1.0	0.0	-1.0	-2.0	-4.0	-3.0	-1.0	-2.0
A_8	-2.0	-3.0	-3.0	-3.0	-3.0	-3.0	-3.0	-4.0	-6.0	-7.0
A_9	0.0	0.0	-1.0	-1.0	-1.0	-1.0	-1.0	-1.0	-1.0	-1.0
A_{10}	0.0	0.0	0.0	0.0	0.0	0.0	0.0	0.0	0.0	0.0
A_{11}	0.0	0.0	0.0	0.0	0.0	0.0	0.0	0.0	0.0	0.0
A_{12}	0.0	0.0	0.0	0.0	0.0	0.0	0.0	0.0	0.0	0.0
A_{13}	0.0	0.0	0.0	0.0	0.0	0.0	0.0	0.0	0.0	0.0
A_{14}	0.0	0.0	0.0	0.0	0.0	0.0	0.0	0.0	0.0	0.0
A_{15}	0.0	0.0	0.0	0.0	0.0	0.0	0.0	0.0	0.0	0.0

Cumulative Profit Coefficients

	TOTAL HANDS PLAYED									
	5,500	6,000	6,500	7,000	7,500	8,000	8,500	9,000	9,500	10,000
A_1	5.0	4.5	3.5	-2.0	3.5	2.5	15.0	5.0	23.5	28.5
A_2	44.0	45.5	40.0	50.0	37.0	21.0	34.5	38.0	42.5	38.5
A_3	-0.5	-3.5	-2.0	-3.0	-13.0	-8.0	-9.5	-4.5	4.5	-3.5
A_4	48.0	43.0	42.5	39.5	37.5	42.0	39.0	33.0	33.5	39.5
A_5	-2.5	2.0	0.0	-1.0	0.0	0.0	-1.5	-4.5	-1.5	-0.5
A_6	8.0	8.5	9.5	9.5	10.0	6.0	8.0	7.0	1.0	-1.0
A_7	0.0	2.5	2.0	3.0	2.0	2.0	2.0	2.0	2.0	3.0
A_8	-6.5	-6.5	-7.5	-6.0	-6.0	-6.0	-5.0	-5.0	-5.0	-3.0
A_9	-2.0	-1.0	-1.0	-2.0	-2.0	-2.0	-1.0	-1.0	-1.0	0.0
A_{10}	0.0	-1.0	-1.0	-1.0	-1.0	-1.0	0.0	0.0	0.0	-1.0
A_{11}	0.0	0.0	0.0	0.0	0.0	0.0	1.0	1.0	1.0	1.0
A_{12}	0.0	0.0	0.0	0.0	0.0	0.0	1.0	1.0	1.0	1.0
A_{13}	0.0	0.0	0.0	0.0	0.0	0.0	-1.0	-1.0	-1.0	-1.0
A_{14}	0.0	0.0	0.0	0.0	0.0	0.0	0.0	0.0	0.0	0.0
A_{15}	0.0	0.0	0.0	0.0	0.0	0.0	0.0	0.0	0.0	0.0

Cumulative Profit Coefficients (continued)

| | TOTAL HANDS PLAYED | | | | | | | | | |
	10,500	11,000	11,500	12,000	12,500	13,000	13,500	14,000	14,500	15,000
A_1	30.5	36.5	40.0	51.5	46.5	62.5	64.5	55.5	19.5	29.5
A_2	43.5	37.0	44.0	55.5	57.0	75.0	82.0	89.5	94.0	108.0
A_3	6.5	5.0	7.0	9.0	10.5	0.0	1.5	-13.5	-23.0	-19.0
A_4	40.5	31.5	41.0	36.0	33.5	35.5	35.5	40.5	40.5	35.5
A_5	3.0	3.0	0.0	0.5	1.5	-2.0	-10.0	-8.0	-6.0	-7.0
A_6	-1.0	-2.0	-0.5	-3.5	-4.5	-5.0	-5.0	-4.0	-3.0	-3.0
A_7	4.0	4.0	3.0	3.0	2.0	4.0	4.0	4.0	-2.0	0.0
A_8	-1.0	-3.0	-4.0	-2.0	-2.0	0.0	0.0	-1.0	-1.0	-1.0
A_9	1.0	1.0	1.0	1.0	1.0	2.0	2.0	2.0	2.0	2.0
A_{10}	-4.0	4.0	-4.0	-4.0	-4.0	-5.0	-5.0	-5.0	-5.0	-5.0
A_{11}	1.0	1.0	1.0	1.0	1.0	1.0	1.0	1.0	1.0	1.0
A_{12}	1.0	1.0	1.0	1.0	1.0	1.0	1.0	1.0	1.0	1.0
A_{13}	-1.0	-1.0	-1.0	-1.0	-1.0	-1.0	-1.0	-1.0	-1.0	-1.0
A_{14}	0.0	0.0	0.0	0.0	0.0	0.0	0.0	0.0	0.0	0.0
A_{15}	0.0	0.0	0.0	0.0	0.0	0.0	0.0	0.0	0.0	0.0

Cumulative Profit Coefficients (continued)

	TOTAL HANDS PLAYED									
	15,500	16,000	16,500	17,000	17,500	18,000	18,500	19,000	19,500	20,000
A_1	22.0	40.0	39.0	44.5	60.5	51.5	46.5	27.5	14.5	18.0
A_2	115.5	133.0	139.0	154.0	153.5	158.5	169.0	167.5	141.5	133.0
A_3	-19.5	-25.5	-30.0	-26.0	-27.5	-17.5	-13.0	-17.5	-16.5	-30.5
A_4	31.5	26.5	28.0	26.0	22.0	23.5	21.5	21.5	18.5	17.5
A_5	-10.0	-9.0	-10.5	-5.5	-6.5	-3.5	-0.5	-1.0	-2.0	-5.0
A_6	-4.0	-4.5	-7.5	-4.5	-6.5	-9.0	-9.0	-11.0	-12.0	-12.0
A_7	0.0	1.0	1.0	2.0	2.0	3.0	0.0	0.0	0.0	0.0
A_8	-1.0	0.0	0.0	0.0	0.0	2.0	4.0	4.0	4.0	4.0
A_9	2.0	1.0	1.0	2.0	2.0	3.0	4.0	4.0	4.0	4.0
A_{10}	-5.0	-5.0	-5.0	-4.0	-4.0	-5.0	-4.0	-4.0	-4.0	-4.0
A_{11}	1.0	1.0	1.0	3.0	3.0	3.0	4.0	4.0	4.0	4.0
A_{12}	1.0	1.0	1.0	2.0	2.0	2.0	3.0	3.0	3.0	3.0
A_{13}	-1.0	-1.0	-1.0	0.0	0.0	0.0	1.0	1.0	1.0	1.0
A_{14}	0.0	0.0	0.0	1.0	1.0	1.0	2.0	2.0	2.0	2.0
A_{15}	0.0	0.0	0.0	1.0	1.0	1.0	3.0	3.0	3.0	3.0
A_{16}	0.0	0.0	0.0	-1.0	-1.0	-1.0	-1.0	-1.0	-1.0	-1.0

Cumulative Profit Coefficients (continued)

	TOTAL HANDS PLAYED									
	500	1,000	1,500	2,000	2,500	3,000	3,500	4,000	4,500	5,000
B_1	15.0	21.0	24.0	22.0	18.0	20.0	9.5	22.0	15.5	9.0
B_2	6.5	0.5	-2.0	-1.0	5.0	11.0	16.5	14.5	17.5	15.5
B_3	-13.5	-12.0	-10.0	-15.0	-17.0	-19.5	-19.5	-23.5	-14.0	-10.0
B_4	2.0	2.0	11.0	8.5	-0.5	-4.5	-4.0	5.0	4.5	8.5
B_5	-2.0	1.0	-2.5	-4.5	-2.5	-5.0	-1.0	2.5	7.0	12.0
B_6	3.0	5.5	8.5	9.5	6.5	5.5	11.5	15.0	13.5	11.0
B_7	-3.0	0.0	0.0	0.0	0.0	1.0	1.0	-2.0	0.0	2.0
B_8	0.0	0.0	-2.0	-3.0	-3.0	-2.0	-4.0	-2.0	-3.0	-4.0
B_9	0.0	0.0	0.0	0.0	0.0	1.0	2.0	4.0	2.5	3.5
B_{10}	0.0	-1.0	-1.0	-1.0	-1.0	-2.0	-1.0	-1.0	-1.0	0.0
B_{11}	0.0	0.0	0.0	0.0	0.0	0.0	1.0	0.0	0.0	1.0
B_{12}	0.0	0.0	0.0	0.0	0.0	0.0	1.0	1.0	1.0	0.0
B_{13}	0.0	0.0	0.0	0.0	0.0	0.0	1.0	1.0	1.0	1.0
B_{14}	0.0	0.0	0.0	0.0	0.0	0.0	-1.0	-1.0	-1.0	-1.0
B_{15}	0.0	0.0	0.0	0.0	0.0	0.0	0.0	0.0	0.0	0.0

Cumulative Profit Coefficients (continued)

	TOTAL HANDS PLAYED									
	5,500	6,000	6,500	7,000	7,500	8,000	8,500	9,000	9,500	10,000
B_1	18.0	32.5	53.0	57.0	53.0	61.5	66.0	47.0	45.5	67.5
B_2	15.5	-1.5	-9.0	-32.0	-15.0	-18.0	-30.5	-48.0	-43.5	-38.0
B_3	-6.0	-5.0	-10.5	-9.0	-2.0	-6.0	2.0	-4.5	-2.0	1.5
B_4	9.0	9.5	7.5	3.0	4.5	1.0	3.5	5.5	-4.5	-2.0
B_5	7.0	11.0	11.0	12.0	10.5	11.5	15.5	11.5	12.0	19.5
B_6	11.0	9.0	11.0	10.0	10.0	8.0	13.0	12.0	10.5	12.5
B_7	2.0	-1.0	-4.5	-4.5	-5.5	-6.5	-7.5	-7.5	-8.5	-4.5
B_8	-3.0	-4.0	-4.0	-4.0	-5.5	-5.5	-6.5	-6.5	-6.5	-5.5
B_9	4.5	4.5	4.5	4.5	4.5	4.5	5.5	5.5	5.5	6.5
B_{10}	1.0	1.0	1.0	1.0	1.0	1.0	0.0	0.0	0.0	0.0
B_{11}	-1.0	-1.0	-1.0	-1.0	-1.0	-1.0	-1.0	-1.0	-1.0	0.0
B_{12}	0.0	0.0	0.0	0.0	0.0	0.0	0.0	0.0	0.0	0.0
B_{13}	1.0	1.0	1.0	1.0	1.0	1.0	1.0	1.0	1.0	1.0
B_{14}	-1.0	-1.0	-1.0	-1.0	-1.0	-1.0	-1.0	-1.0	-1.0	-1.0
B_{15}	0.0	0.0	0.0	0.0	0.0	0.0	0.0	0.0	0.0	0.0

Cumulative Profit Coefficients (continued)

	10,500	11,000	11,500	12,000	12,500	13,000	13,500	14,000	14,500	15,000
B_1	88.0	83.5	87.0	91.0	88.0	106.0	95.0	77.0	53.5	47.0
B_2	-34.5	-39.0	-30.5	-33.0	-57.0	-62.0	-62.0	-57.5	-52.5	-34.0
B_3	1.5	11.5	16.0	19.5	12.5	8.0	11.0	10.5	24.5	28.5
B_4	-3.5	-10.5	-10.0	-4.5	-7.5	-7.5	-8.0	-11.5	-20.0	-16.5
B_5	18.0	17.0	17.5	20.5	16.5	13.0	14.5	11.0	8.0	7.0
B_6	13.5	12.5	13.0	13.5	13.5	13.0	12.5	11.5	9.5	9.0
B_7	-4.5	-7.5	-6.5	-7.0	-7.0	-8.0	-7.0	-7.0	-8.0	-9.0
B_8	-4.5	-4.5	-4.5	-6.5	-6.5	-6.5	-9.5	-9.5	-9.5	-8.5
B_9	7.5	7.5	8.5	8.5	8.5	8.5	8.5	8.5	8.5	11.5
B_{10}	1.0	2.0	2.0	2.0	2.0	2.0	2.0	2.0	2.0	1.0
B_{11}	1.0	1.0	0.0	2.0	0.0	-2.0	0.0	0.0	0.0	0.0
B_{12}	-2.0	-2.0	-2.0	-2.0	-2.0	-4.0	-2.0	-2.0	-2.0	-2.0
B_{13}	1.0	1.0	1.0	1.0	1.0	1.0	1.0	1.0	1.0	1.0
B_{14}	-1.0	-1.0	-1.0	-1.0	-1.0	-1.0	-1.0	-1.0	-1.0	-1.0
B_{15}	0.0	0.0	0.0	0.0	0.0	0.0	0.0	0.0	0.0	0.0

TOTAL HANDS PLAYED

Cumulative Profit Coefficients (continued)

				TOTAL HANDS PLAYED						
	15,500	16,000	16,500	17,000	17,500	18,000	18,500	19,000	19,500	20,000
B_1	77.0	86.5	96.5	75.5	64.5	68.5	50.5	41.5	58.0	52.5
B_2	-52.0	-73.5	-91.5	-87.0	-70.5	-66.0	-68.0	-69.0	-67.5	-60.0
B_3	22.0	36.0	34.5	34.0	29.5	33.5	40.5	32.5	25.5	28.0
B_4	-12.5	-6.0	-2.5	-0.5	-8.0	-6.5	1.5	5.0	11.0	13.5
B_5	0.5	1.0	4.0	4.5	1.5	0.5	-3.5	-5.0	-6.5	-7.5
B_6	8.0	11.0	16.0	12.0	10.0	12.0	14.0	13.0	8.0	8.0
B_7	-9.0	-4.0	-7.5	-6.5	-6.5	-10.5	-9.0	-9.0	-11.0	-14.5
B_8	-8.5	-13.0	-11.0	-10.0	-10.0	-10.0	-11.0	-12.0	-12.0	-12.0
B_9	11.5	12.5	14.5	15.5	15.5	15.5	16.5	16.5	16.5	16.5
B_{10}	1.0	-1.0	-1.0	-2.0	-2.0	-2.0	-1.0	-1.0	-1.0	-1.0
B_{11}	0.0	0.0	-1.0	-1.0	-1.0	-1.0	-2.0	-2.0	-2.0	-2.0
B_{12}	-2.0	-2.0	-2.0	-2.0	-2.0	-2.0	-2.0	-2.0	-2.0	-2.0
B_{13}	1.0	1.0	1.0	1.0	1.0	1.0	1.0	1.0	1.0	1.0
B_{14}	-1.0	-1.0	-1.0	-1.0	-1.0	-1.0	-1.0	-1.0	-1.0	-1.0
B_{15}	0.0	0.0	0.0	0.0	0.0	0.0	0.0	0.0	0.0	0.0

Cumulative Profit Coefficients (continued)

To make use of the profit coefficients (As and Bs), you merely plug them into the simple formulas (I have given them the aesthetically appealing name of "profit equations") presented below:

$$W = X_1A_1 + X_2A_2 + \ldots \ldots X_nA_n$$
$$L = Y_1B_1 + Y_2B_2 + \ldots \ldots \ldots Y_nB_n$$
$$\textbf{Profit} = \textbf{W} - \textbf{L}$$

Where:

X_1 = bet size after single win
X_2 = bet size after two wins in a row
.
.
.
X_n = bet size after n wins in a row

Y_1 = bet size after a single loss
Y_2 = bet size after two losses in a row
.
.
.
Y_n = bet size after n losses in a row

As an example, suppose you were testing the double-up system. Then, you might bet one dollar after every win and double up after every loss. Your X and Y bets would be:

$X_1 = X_2 = X_3 = \ldots X_n = \$1.00$
$Y_1 = \$2, \; Y_2 = \$4, \; Y_3 = \$8, \; Y_4 = \$16$, etc.

To compute the profit after 500 hands, use the corresponding columns from the table to get the As and Bs. These are given as:

$$A_1 = 6.0 \qquad\qquad B_1 = 15.0$$
$$A_2 = 8.0 \qquad\qquad B_2 = 6.5$$
$$A_3 = 12.5 \qquad\qquad B_3 = -13.5$$
$$A_4 = 7.0 \qquad\qquad B_4 = 2.0$$
$$A_5 = 6.5 \qquad\qquad B_5 = -2.0$$
$$A_6 = -2.0 \qquad\qquad B_6 = 3.0$$
$$A_7 = 1.0 \qquad\qquad B_7 = -3.0$$
$$A_8 = -2.0 \qquad\qquad B_8 = B_9 = \ldots B_n = 0$$
$$A_9 = A_{10} = \ldots A_n = 0$$

Then:

$$W = 1(6) + 1(8) + 1(12.5) + 1(7) + 1(6.5) + 1(-2)$$
$$+ 1(1) + 1(-2) = 37$$
$$L = 2(15) + 4(6.5) + 8(-13.5) + 16(2) + 32(-2) + 64(3)$$
$$+ 128 (-3) + 0 = -276.$$
$$\text{PROFIT} = W - L = 37 - (276) = -239$$
$$\text{PROFIT} = -\$239.00$$

It should be emphasized that the profit coefficients are cumulative. That is, if you used the As and Bs from the twenty-thousand-hand column, for example, you would get the profit for the entire twenty thousand hands. You would not get the results for the 500 hands between 19,500 and twenty thousand hands only. So, you can obtain the cumulative result for intermediate hand totals by taking the As and Bs from the appropriate columns in the table. A plot of cumulative profit versus total number of hands can be very quickly generated using this information.

If you have some run- or sequence-based system in mind, give the profit coefficients a whirl. You can accomplish in a few minutes what it took us months to do. You will be surprised at the way your profit will fluctuate, but it is better to find out this way than in the casino.

If you still have some questions about runs, the use of the profit coefficients, or the way in which they were developed, you can find more detailed information in chapter 7. If you wish to pursue the

subject even more deeply than this, equations are derived in Appendixes B and C for determining the profit coefficients from run
statistics.[2]

It is critically important that you know that this set of basic
strategy hands turned out to be quite favorable to the player. Instead
of an expected +0.1 per cent player's edge, we got about +0.6 per cent
for these twenty thousand hands. Bear this constantly in mind for
any tests you might perform, and in the ensuing discussions. Any
results you get will tend to be optimistic. If you got bad results here,
you will probably do a lot worse in the long run. If, on the other
hand, you get favorable results, you might consider extending the
sample size using the methods in chapter 7. Such an extension could
provide hours of amusement and may save you a lot of money.

Finally, this particular data set was not chosen from a vast quantity of data. Rather, we simply played twenty thousand hands and
recorded the results with no selection whatsoever. This is just what
came out of the hopper on the first draw, nothing more.

In subsequent discussions, let us agree to call these particular
twenty thousand hands "the data package."

THE FLAT BET SYSTEM

This system will provide the best baseline for comparing the
results of other systems played through the data package. The
betting rule is merely:

Flat bet one unit on all hands.

If you will look at the solid line in the following graph, you will
find a profit for a five-dollar flat bet. The five-dollar betting level was
chosen arbitrarily.

As you can see, these data are slightly favorable to the flat bettor.
It shows a gradual, albeit fluctuating rise in capital over the entire
twenty thousand hands. Therefore, the data package clearly represents a series that is moderately biased in favor of the player.

From this, we may at least hypothesize:

If a betting system does not show a favorable history over this data package, it is probably a very poor investment.

THE "GENERAL MIN-MAX" SYSTEM

Also shown on the graph is a significantly better profit history depicted by the light line. The betting rules for this system are:

1. **Place *your* maximum bet after all wins.**
2. **Bet the table minimum after all losses.**
3. **Ignore pushes.**

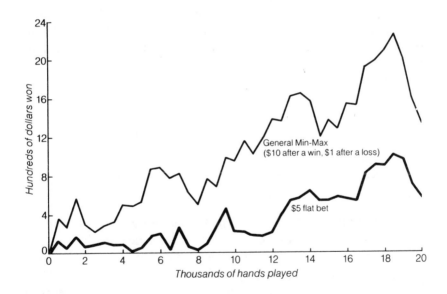

The Flat Betting and "General Min-Max" Systems Test Results

I call this system "General Min-Max" because it is almost identical to my "Special Min-Max System" of chapter 5, but it is more general. If you add pushes to rule 1, and four-card hands to rule 2, then you will have the chapter 5 Min-Max System. This is definitely one of my favorite betting techniques. In fact, I devised and used it successfully long before making the discoveries of chapters 4 and 5. Surprisingly enough, I have not seen this simple scheme published anywhere else.

The history in the graph is based on a ten-dollar maximum and a one-dollar minimum bet. At these levels, your total investment for the twenty thousand hands would be about equivalent to that for the five-dollar flat bet. The basic improvement in profit is due to better money management. That is, *you win at your maximum bet level during winning sequences and lose at the table minimum during losing sequences*. Since the data package is a favorable one overall, the winning sequences are frequent and lengthy enough to make it profitable to use this system.

Can the General Min-Max System lose during less favorable sessions? Yes, it certainly can. Look at the last fifteen hundred hands on the graph, for example. The General Min-Max scheme lost some $920.00 while the flat bet was losing only $440.00 on a roughly equivalent investment.

Applications to Roulette, Craps and Other Card Games

If you are going to play blackjack, you can use the methods of chapter 5 just as easily as the General Min-Max idea. In this way, you gain a favorable edge on the house while incorporating good inherent money management principles.

The General Min-Max concept also has merit as a betting system for any game with an essentially even money payoff. Now, if the house has the edge, and except for blackjack it invariably does, you are not going to win in the long run with any system. However, if you wish to play another game for fun, and hope for at least a slightly favorable short run session, then you will need a good plan for maximizing your wins and keeping your losses to a minimum. I would recommend General Min-Max betting in this case.

ROULETTE. A typical application might be red-black betting at roulette, which is an even-money game. You might start with a one-unit bet on black, for example. If you win, bet say, two units (or some other maximum) on black and stay at this level until you get a loss. Bet one unit after all losses. In this way, you will certainly last a very long time on even a small amount of capital. Moreover, if you have a favorable session, you could get ahead a few dollars and quit winners. Bear in mind, though, that you are trying to overcome a basic house edge of 5.26 per cent on a double zero wheel or 2.70 per cent for a single zero wheel.[3] If you are lucky enough to accrue a little profit at roulette, be smart and quit before you lose your way back into your own pocket.

CRAPS. I would particularly recommend this General Min-Max approach if you are a smart craps player. By simply playing the pass line, you have only a 1.4 per cent house edge to overcome. Wilson[4] and others have demonstrated that this edge may be reduced to as little as 0.848 per cent by "taking the odds." In some casinos, the "don't pass" bettor can reduce this edge to about 0.5 per cent by "laying the odds." This house advantage is so small that, for all practical purposes, it will be inundated by statistical fluctuations unless you play for many thousands of trials. The simple Min-Max technique will give you a decent shot at winning if you get situated at a moderately hot table.

There is a hooker in the Min-Max scheme of course, or it could never lose as it did in the last fifteen hundred hands of the data package. An unusually large concentration of sequences that oscillate in the fashion of win, lose, win, lose, win, lose (ad nauseam), will do you in. But do not worry about it too much because such sequences are rarely very long or highly concentrated. If they were, then the Martingale-type systems would invariably win. The best medicine for curing any gnawing doubts is to knock out some hands for yourself and see how it goes.

For the Real Swinger

If you are a "high roller," you will be interested in the next graph. The upper curve is for the General Min-Max System with a $500 maximum and a five-dollar table minimum bet. This represents about as well as you can do on this data package (except for the chapter 5 and card counting methods, of course). The graph shows a profit peak of about $114,000 at the 18,500-hand mark, and a $47,700 losing streak during the last fifteen-hundred hands!

This striking record gives a clear indication of the profits that could conceivably be racked up with this simple betting method. But can you afford a fifty thousand dollar losing streak or worse? If you can, have at it with my blessing.

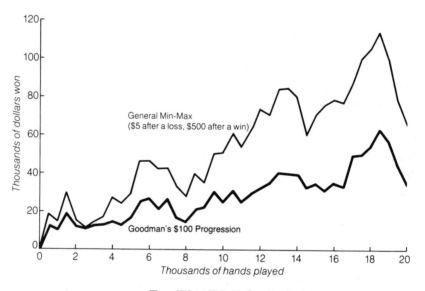

For The Real Swinger

The lower line on the graph shows a similar profit history but of lesser magnitude. This figure is based on the riskiest of Mike Goodman's[5] recommended progressions. These are investigated next.

GOODMAN'S PROGRESSIONS

> *When betting someone else's money be brave, when betting your own be a coward.*[6]
>
> MIKE GOODMAN, 1963

Personally, I found Mike Goodman's book colorful and fascinating to read, although somewhat inaccurate relative to blackjack. His recommended playing strategy is close to that of chapter 3, but incorrect in many instances. We once played ten thousand hands with his strategy and got about +0.4 per cent for the player. I would guess that a more accurate figure would be in the –0.5 per cent to –1.0 per cent bracket, however.

His section on "Money Management in 21"[7] is of considerable interest to the system player, nevertheless. In fact, I would consider it a major contribution to the world of systematic betting. What he proposes, in capital letters, is this:

YOU MUST PROGRESS AS YOU WIN.

To elaborate, pick some basic betting level, say two dollars, that appeals to your budget. If you *lose* a hand, *stay* at this level. If you *win* a series of hands, *increase* your bet a little after each *win*. He suggests you use a winning betting sequence of two, four, six, ten, ten, fifteen, fifteen, fifteen . . . fifteen dollars, if you are a "two-dollar bettor." In his words you are "leaving a little, taking a little"[8] or employing a "slow progression." The most interesting feature is the progression to some maximum value that is tailored to both your budgetary and psychological limitations.

The astute casino executive and pit boss also makes this interesting statement, "Once you start to win a few hands in a row you are now playing with someone else's money—the club's—and I assure you they won't like it one bit."[9] While the psychology here is appealing, be assured that you are playing with your own money at all times. Subsequent discussion will show you just how much capital may be required to support such a betting system, even for the two-dollar bettor; which is certainly *your* problem and not the club's. Notwithstanding, his ideas are basically sound. You will definitely be a "tough" bettor if you follow his advice.

His recommended progressions are summarized in the following table:

			DURING WINNING SEQUENCES BET						
BETTING CLASS	AFTER EACH LOSS BET	After 1st Win	After 2nd Win	After 3rd Win	After 4th Win	After 5th Win	After 6th Win	After 7 or more Wins
$2	$2	$4	$6	$10	$10	$15	$15	$15
$5	$5	$10	$15	$25	$25	$35	$50	$50
$10	$10	$20	$30	$50	$50	$75	$100	$100
$25	$25	$50	$75	$125	$125	$150	$200	$200
$50	$50	$100	$150	$225	$225	$300	$300	$300
$100	$100	$200	$300	$400	$400	$500	$500	$500

Goodman's Progressions

The most ambitious of these, the "$100.00 bettor" progression, is represented by the lower line of the previous graph. As you can see, it gives a similar but less profitable record to the General Min-Max history also shown on that graph. So, if you are that big a spender, you might as well go to the Min-Max technique and go for all the marbles.

Actually, Goodman's progressions can be considered a special case of the General Min-Max approach. A Min-Max betting set can be found that is roughly equivalent to any of Goodman's progressions. For example, the General Min-Max system with a seven-dollar maximum and one-dollar minimum gives an almost identical profit trace to Goodman's two-dollar progression over the twenty thousand-hand data package.

Therefore, with Goodman's two-dollar system you should not consider playing with less than $700 to $1,000 backing you up. You can feel pretty safe with the two-dollar maximum and one-dollar minimum General Min-Max System and a $100 to $200 bankroll, on the other hand.

I ran across an almost identical recommendation to Goodman's progressions in a little paperback book published by Gambling International[10] and endorsed by "Nick the Greek" Dondolas and, so the booklet claims, "the world's greatest gambler," Duke Spartan. A playing strategy similar to Goodman's, and containing similar errors, is proposed along with a positive progression betting technique.

The section of real interest to us was written by Ben Holiday, a close friend of Duke Spartan, and entitled "Money Control." He introduces the concept of establishing a minimum bet, which he calls your "Wagering Conceit," based on one-twentieth of your bankroll. You then bet this "wagering conceit" after all losses. For a one-dollar minimum he proceeds to recommend a betting progression of one dollar, two dollars, three dollars, five dollars, ten dollars . . . ten dollars during winning runs. This is obviously quite similar to the Goodman two-dollar progression.

A General Min-Max ratio of four or five to one would yield about the same results. The wagering conceit of one-twentieth of your capital would give you a minimum chance of ultimate ruin of some 80 per cent. In fact, with the increased betting level during a series of wins, your chance of ruin is in excess of 90 per cent. You had better refer to chapter 6 and size your bets and bankroll to a safer level unless you find this extremely risky outlook to your satisfaction.

One other point will be raised before we leave Goodman's progressions for a quick look at the "Oklahoma" system. That is, his progressions could be more profitable if all bets following losses were made at the table minimum. If the purpose of maintaining a flat bet during losing sequences is to minimize losses and keep out of trouble, then a more effective job would be accomplished by betting the table minimum during these sequences. An analysis of the data package and profit coefficients bears this out.

THE OKLAHOMA SYSTEM

This technique is merely another kind of special case of General Min-Max-type betting which I discovered in a book by Sidney

Radner.[11] While this particular system has value, I would be a little wary of his advice, especially on blackjack. To his credit, he has described a goodly number of interesting, if unprofitable, systems that can be applied to other casino games as well as blackjack.

The "Oklahoma" system got its name, according to Radner, "because a gambler from that state presumably tried it out and found that it worked." The betting rules are similar to Goodman's, but with a new twist:

1. Bet one unit after all losses.
2. Bet two units after the first win and four units after the second win.
3. Win or lose, go back to one unit after the next bet and start over again.

The main difference between this and say, General Min-Max or Goodman's progressions, is the recycling after a couple of wins. In any event, it is almost identical to General Min-Max with a three-unit maximum and does equally well on the data package. Its profit history looks very much like those in the two graphs previously discussed with an appropriate scale change.

This ends our discussions of what we might term "positive progressions." Most of the remaining systems fall into a general category we can call "negative progressions." This is a pretty good name for a couple of reasons. First, the subsequent systems are based on increasing your bet while you are on a losing streak and second, they mostly all lose drastically on the data package. Read on and avoid some of the biggest and most common pitfalls that confront the gambling community.

THE MARTINGALE SYSTEMS

Small Martingale (Double-up)

This particular technique has got to be the most popular gambling system known to mankind. So many people with whom I have discussed my work on blackjack have said in effect, "I too have a great idea for a blackjack system. I don't see how it can lose." They then proceed to describe the same old "double-up" or small Martingale technique as if it were a brilliant breakthrough.

Frankly, I am sick and tired of hearing about it and wish to, once and for all, debunk the abhorrent concept and banish it forever from the minds of intelligent gamblers; *especially* informed blackjack players. On this particular subject, I am undeniably adamant.

The small Martingale is a lousy, impractical and disastrous system to use in a casino.

Sure, in theory, if you had unlimited capital, and if you could find a casino that would let you bet any old amount that you wanted, then you would have to win. Sure, and if you had wings you could fly, too. The simple blaring truth is, however, that every casino in the world has a house maximum bet. Ergo, the "double-up" is automatically defeated. But we do not stop here. Let us take a good hard look at this technique in action. Let us give it the coup de grace.

With this betting technique you double your bet after each loss. During winning runs, on the other hand, you hold your bet constant. Say, for example, your basic bet is one dollar. A typical series of bets would be as follows:

Hand Number	Outcome	Bet
1	Win	$1
2	Win	$1
3	Win	$1
4	Lose	$2
5	Lose	$4
6	Win	$1
7	Lose	$2
8	Lose	$4
9	Lose	$8
10	Lose	$16
11	Win	$1
12	Win	$1

After each losing series you gain one unit. To see this, look at hands four, five, and six. On hand number four, one dollar was lost. Two dollars were lost on hand number five and four dollars were won on the sixth hand. This tallies up as follows:

Hand Number	Outcome	Dollars Won	Dollars Lost
4	Lose		$1
5	Lose		$2
6	Win	$4	

TOTAL WON = $4 – $3 = $1

One dollar was also won at the conclusion of the losing series starting with hand number seven. This series goes as follows:

Hand Number	Outcome	Dollars Won	Dollars Lost
7	Lose		$1
8	Lose		$2
9	Lose		$4
10	Lose		$8
11	Win	$16	

TOTAL WON = $16 – $15 = $1

Of course, this system is impractical since it will be stopped by the table maximum. Suppose, for example, the table maximum was fifty dollars. You would bet, for each long losing series, one, two, four, eight, sixteen, and thirty-two dollars. The double-up system would require that the next bet be sixty-four dollars, but since the maximum allowable bet is fifty dollars, it could not be placed. The system is thus defeated.

Even if there were no maximum, you would have to be very rich to play this system. During the twenty thousand hands of the data package, many long losing streaks came up. The longest of these was a run of fourteen, occuring at about three thousand hands.

For this run, the bet would be as follows:

Outcome	Amount Lost
Win	
Lose	$ 1
Lose	$ 2
Lose	$ 4
Lose	$ 8
Lose	$ 16
Lose	$ 32
Lose	$ 64
Lose	$ 128
Lose	$ 256
Lose	$ 512
Lose	$ 1024
Lose	$ 2048
Lose	$ 4096
Lose	$ 8192

TOTAL LOST = $16,383

So you see, you would have needed $16,383 to cover your losses. Moreover, an additional $16,384 would have been required to place the next bet. The grand total needed to win one dollar would be a heart stopping $32,767.

There were many other lengthy losing runs of eight to thirteen losses in a row, but the profit history really rams the message home. Take a peek at the next graph. Here we have the profit record over the twenty thousand-hand data package for the small Martingale, giving it every possible advantage. The house maximum was set at $500 and the table minimum at one dollar. This range is rarely found at a blackjack table. In fact, I have never seen a table this liberal. The typical minimum for the $500.00 maximum game is five dollars. Moreover, and this should be highly stressed, this data package represents *very favorable* data for the player! Refer back to the first graph in this chapter and refresh your memory as to the healthy flat bet advantage that the package provides.

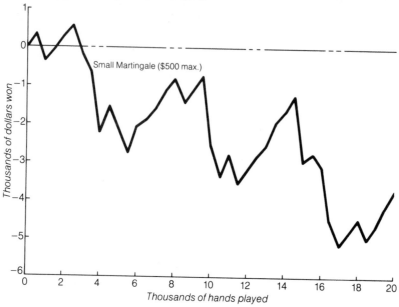

The Small Martingale or "Double-Up" System Test Results

Okay, returning to the present graph, we can see a clear, steady, depressingly steep drop in capital as the game wears on. By the time three thousand hands have gone by, the system is in the red to stay. This record produces the disgusting loss rate of some twenty-five cents a hand. At roughly 100 hands an hour, this wonderful method would cost you an estimated $25.00 for every hour that you played it! This is probably a conservative estimate too, since the data are biased for the player.

The case is nearly as bad for the table with a $50.00 maximum as you can see in the next graph. If you modify the system and remain at $50.00 until you win a hand, once this level has been reached, you will have the system exemplified by the upper line in the graph. As you can see, the modified method is little better, leaving the player in the red some eighteen thousand hands out of twenty thousand.

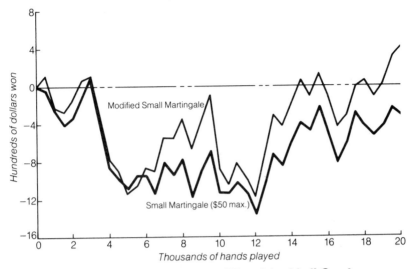

The Small Martingale or "Double-Up" System
Test Results

Now I ask you, is this a good way to bet? Do you want to invest your money in a loser like this? By now you should be convinced of the danger of such a path. Never use the Small Martingale system. Doubling up will break your heart and your bankroll if you are foolish enough to listen to the multitude of self-styled experts who expound this type of garbage.

Grand Martingale

The Grand Martingale system is a variation of the Small Martingale just discussed. With this variation, you also hold the bet constant after each win. After each loss, however, instead of merely doubling the bet, you *double up and add one.*

That is, if your first loss was $1.00, you would bet $2.00 + $1.00 = $3.00 on the next play. If you lost that hand as well, you would then bet $6.00 + $1.00 = $7.00. After still another loss, you would bet $14.00 + $1.00 = $15.00 and so on as the losses piled up.

It will next be shown that this system will produce a win of *one unit for every hand played.* Assume you had a hypothetical series that went as shown:

Hand Number	Outcome	Amount Bet	Amount Won	Amount Lost
1	Win	$1	$1	
2	Lose	$1		$1
3	Lose	$3		$3
4	Lose	$7		$7
5	Win	$15	$15	

TOTAL WINNINGS = $16.00   TOTAL LOST = $11.00
TOTAL WON FOR FIVE HANDS = $5.00

You would bet one dollar after the first win, three dollars after the first loss, seven dollars after the second and fifteen dollars after the third loss. You lost a total of $1 + $3 + $7 = $11 on hands two, three and four and you won one dollar on hand number one and fifteen dollars on hand number five. So your total winnings for the five hands is $16 – $11 = $5, which represents one dollar for each hand played.

Naturally, you must pay a price for using the exciting Grand Martingale instead of the Small Martingale. That price is the fact that you will reach the table maximum sooner with the Grand Martingale.

For example, again assume that the maximum allowable bet is fifty dollars. Now, compare the two Martingale systems in a straight losing series for a basic unit of one dollar:

Outcome	Small Martingale Bet	Grand Martingale Bet
Lose	$2	$3
Lose	$4	$7
Lose	$8	$15
Lose	$16	$31
Lose	$32	$63 (over the $50 limit)

You can see that the Grand Martingale reaches the limit *one bet sooner!*

Profit records for the Grand Martingale system based on our data package are shown in the next graph for house betting limits of five hundred dollars and fifty dollars. For the fifty-dollar maximum case, the bet was recycled to one-dollar after a thirty-one-dollar bet level was reached. A five hundred dollar bet was made in place of the required $511 for the $500 maximum situation.

I almost hate to show you this graph and its five-hundred-dollar maximum curve because I feel it is misleading. At first glance it looks like a real winner, does it not? But could you sustain the initial six thousand-hand losing streak? Could you afford the three thousand dollar loss depicted at about ten thousand hands?

Of course, the plot does look good at first, but let us take an even closer look. In the next graph, both profit and hand total scales have been doubled and the first ten thousand hands replotted. The heavy line represents the Grand Martingale with the five-hundred-dollar maximum. Just look at those oscillations in capital. Do you want to sit through something as horrible as this? You would need plenty of confidence and patience to be satisfied to be at –$1,000 after some fifty-five hours of playing!

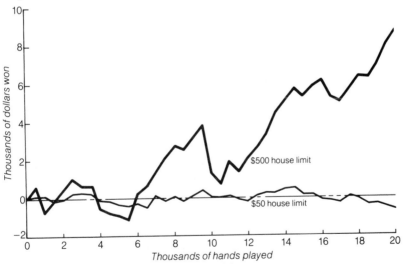

**Full Blown Grand Martingales
Test Results**

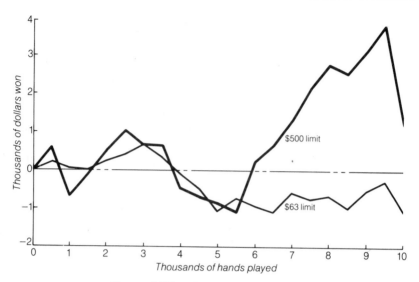

Grand Martingale Systems
Magnification

On the other hand, if you were lucky enough to sit down at say, the fifty-five-hundred-hand point, your profit rise would have been exhilarating, indeed! You would have run up a tidy five thousand dollar gain after some forty hours, or so, of work. But remember, this is a particularly nice sample of data and liable to give optimistic results. I would be *very* cautious about playing this system.

The flag of caution is also raised by the other profit histories recorded for Grand Martingales with lesser maximum bets. For a house limit of fifty dollars (thirty-one-dollar maximum Grand Martingale bet), the graph depicts a typical losing series ending up six hundred dollars in the red after twenty thousand hands. Again, allowing the progression to develop one bet further to a sixty-three-dollar limit, the magnified curve is equally distressing. Here we see a gradual erosion of capital to about minus one thousand dollars after ten thousand hands. Although not graphed, a similar trend continues for the next ten thousand hands also.

What I think we are seeing in the five hundred dollar maximum case is merely another statistical fluctuation; this time apparently favorable to the player. Should we extend the data base considerably, then the five hundred dollar maximum would be *tested* more often and you would see the profit history looking more pessimistic, I am sure.

In summary, any type of Martingale progression is a risky business at best. However, if you wish to build your bankroll very quickly, and you are a real gambler, the full-blown Grand Martingale, together with the basic strategy and an adaptation of the complete system of chapter 5, could conceivably do it if you get a little lucky to boot. Never try this for more than a few hands, though, or it will catch up with you for sure.

Limited Martingales

Radner broaches the subject of "Limited Martingales" in his discussion of roulette systems where he states:

> All Martingales are limited, of course, by the amount that the house will let a player venture on a single turn of the wheel, but with the 'Limited Martingale' the ceiling is set by the player himself, according to the amount he feels that he can afford.[12]

Some "severely limited Martingales" are illustrated by the profit traces in the following graph. The three-dollar limit shows the typically poor Martingale record. A negative profit for nearly the whole twenty thousand hands. But the progression with a seven-dollar limit looks a little better. Another favorable fluctuation? Probably, but the accompanying General Min-Max trace for a maximum bet of four dollars and a minimum of one dollar should be sufficient to sway you away from this particular Martingale anyway. General Min-Max is just so much better! It never went negative, and is generally a much better behaved system. Why get yourself in trouble "chasing your money" with a Martingale, when you can do so much better?

Added weight is given to this statement by the next graph. Here you will find magnified ten thousand-hand records of Martingales with a fifteen- and thirty-one-dollar limit. The typically wild and predominantly negative Martingale trace is once again observed. I rest my case.

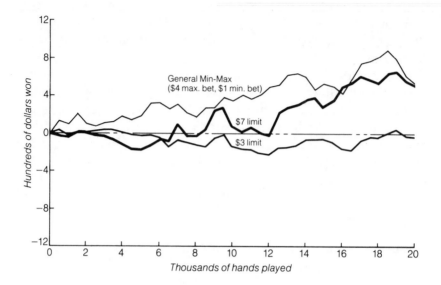

Severely Limited Martingales
Test Results

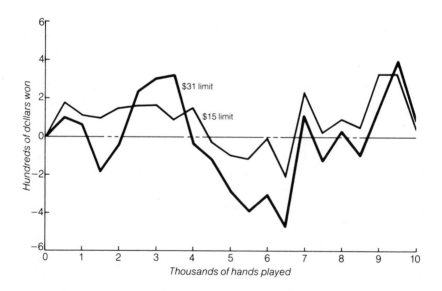

Limited Martingale Systems
Magnification

The Limited Martingale might be better termed a "controlled progression." The so-called "Count of Monte Carlo"[13] has also applied this term to the Labby system, which is treated next. You will learn more about the Count, including my own encounter with him, in the next section.

THE "LABBY," "CONTROLLED PROGRESSION" OR "CANCELLATION" SYSTEM

According to Radner, "The word 'Labby' is supposedly a nickname for a famous journalist, Labouchere, who used it to finance his visit to the Mediterranean's famous rock."[14] Wilson refers to the identical method as the "Cancellation System."[15] While the rationale behind the technique appeals psychologically to the system-playing gambler, it is really just a big fiasco. Especially when applied to blackjack without counting cards.

To play the Labby, you must use pencil and paper at all times. In itself, this is a huge disadvantage in blackjack where your hands are already occupied with both chips and cards. You would really have to team up with someone else to comfortably implement the method at a blackjack table.

First, you must decide how many units you wish to have as your systematic goal. Whatever goal you chose, you break it up into an arbitrary set of smaller units. For example, a six-unit goal could be broken up into a column of numbers as follows:

1
2
2
1

This will initiate your working column. Your first bet will be the sum of the top and bottom numbers on the column, or in this case, 2. If you win, you cross them off and bet the total of the new top and bottom numbers. Your column would look as follows in a series of straight wins.

	Start	After 1st Win	After 2nd Win	
	1	~~1~~	~~1~~	1
	2	2	~~2~~	2
	2	2	~~2~~ Start over	2
	1	~~1~~	~~1~~	1
Bet	1 + 1 = 2	2 + 2 = 4		1 + 1 = 2
Total Won		2 Units	2 + 4 = 6 Units	

After the second win, you have successfully "closed out" your column and won your six units. A new column is started and the procedure is repeated. If you sustain a loss, instead of crossing off the top and bottom numbers you add a new one, totaling these, to the bottom of your column. A series of two losses followed by two wins would look as shown below:

Outcome	Start	Loss	Loss	Win	Win
	1	1	1	~~1~~	~~1~~
	2	2	2	2	~~2~~
Column	2	2	2	2	2
	1	1	1	1	1
		2	2	2	~~2~~
			3	~~3~~	~~3~~
Next Bet	1+1=2	1+2=3	1+3=4	2+2=4	2+1=3
Total Won		−2	−5	−1	+3

As is apparent, a third win would close out the column and you would have won the desired six units. At this point, you start all over again with a new column. In summary, the Labby rules are simply:

1. Break up your goal into a column of smaller units.
2. Always bet the total of the top and bottom numbers.

3. If you win, cross off these two figures.

4. If you lose, add their total to the bottom of the column.

5. If you close out a column, start a new one.

There is a man who calls himself the "Count of Monte Carlo" who has published a booklet on blackjack that highly recommends this system. This same man has published a number of these booklets on various casino gambling games and on betting the dogs and horses as well. He also ran and published an American magazine exclusively devoted to gambling. This was a fairly nice quality magazine.

Allow me to digress from the Labby at this point and I will tell you about my experience with him. Having seen the magazine on the magazine racks in the Cape Kennedy area, I wrote for advertising rates. Based on these, I sent $168 to them several months later to run an ad for the Du-Rite Wheel. Well, I received a letter a short time later from the "Count." He informed me that the magazine was folding up, but he furnished some literature and rates on a new magazine which he said would come out in two or three months.

There was considerable other correspondence between the two of us relative to a personal publicity story he was going to throw in for free about me and my wheel, and other possible business arrangements we might make on the wheel. But to make a long story short, he apparently never published the magazine!

I finally phoned him several times but could not reach him. To my surprise, he did return the call one day, at his expense, and assured me that the magazine would be published within thirty days. Several months and no magazine later, I started a new series of letters again requesting a refund. To his credit, he finally did cough up my refund!

While I found both the Count and his publications to be interesting, his blackjack strategy is riddled with nonsense and, as is next demonstrated, his "Controlled Progression" is particularly poor as a blackjack system.

This routine has tremendous appeal to those who love systems. After all, every time you win you cross off *two* numbers, and when you lose you only add *one* number to your column. Since you cannot lose twice as many times as you win, they reason, then surely you must inevitably close out your column. Got you hooked? Let us take a look at it under the data package, before we get too excited.

A goal of three units was chosen for testing the Labby on the data package. The working column was initiated in this form:

1

2 (three-unit total)

Now this system does not lend itself readily to a "profit coefficients" type of analysis since the bet size is variable during different sequences containing runs of identical length. So, I had the outcomes of all twenty thousand hands coded and punched out on carefully checked IBM cards. I then wrote a simple computer program to play the Labby against the identical hands of the data package. Allowing a house maximum of five hundred dollars and a liberal accompanying table minimum of one dollar, the program reset the Labby column at the initial three units shown above whenever a bet in excess of five hundred dollars was required.

The results are depicted in the following graph. One data point is plotted for every five hundred hands, as before, but every "reset" is clearly marked on the chart. Is not this a terrible profit record on our admittedly "favorable" data? The system required bets in excess of the $500 maximum, and was thus defeated and recycled, some sixteen times during the twenty thousand-hand test. After an initial small profit gain at about five hundred hands, the system was always in the red thereafter, reaching a peak loss of close to fifteen thousand dollars at the thirteen thousand-hand mark.

The reason that the system fails so miserably is this. Although the basic blackjack strategy gives a slightly favorable *percentage,* you actually lose more *hands* than you win. This is because you tend to *win* more of the "extra money" bets than you lose. You win *all* blackjacks, paying 3 to 2, for example. As for splits and double-down hands, naturally you also win more of them than you lose, because that is why the basic strategy was constructed in the first place. So, if you win on this type of hand often, then you achieve your even game in less winning hands than losing hands. My computer run shows that out of five million hands, the numbers of losing, winning, push and zero-effect hands were as follows:

TOTAL WINNING HANDS   2,183,283
TOTAL LOSING HANDS   2,396,297
TOTAL ZERO-EFFECT HANDS   14,206
TOTAL PUSHES   406,214

Zero-effect hands are splits in which you lose one hand and win the other.

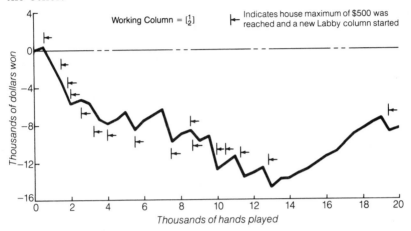

The "Labby" or Controlled Progression

So, in the case of the Labby, you will close out columns until you start to get an unfavorable series of hands. Then your column will become larger and larger. Moreover, the figures in it begin to grow at an alarming rate. For example, a new column starting off with four losses and two wins would develop like this:

START	LOSE	LOSE	LOSE	LOSE	WIN	WIN
1	1	1	1	1	1̸	1̸
2	2	2	2	2	2	2̸
	3	3	3	3	3	3
		4	4	4	4	4
			5	5	5	5̸
				6	6̸	6̸

Now look at the top and bottom figures remaining in your column. The top figure has grown to a three and the bottom to a four. If you now start a new losing streak, your bets will increase by three units every time. So, three more losses would require bets of seven, ten, and thirteen units, respectively. In fact,

Every time you win without closing out a column, your betting rate increases.

This so-called "controlled progression" very rapidly gets out of control! So rapidly, in fact, that you will run into the house maximum at least as often as depicted on the graph.

A proposed remedy to this problem is to break up the column when the house maximum is reached. The argument is that you can proceed to close out a new column *individually* without hitting the maximum. What utter nonsense! If the system does not work for your first column, it is certainly not going to work on a flock of new ones! Take another look at the graph and see for yourself that starting over does not help.

Before we leave the Labby, I would like to make one final point. The aforementioned "Count" also says that you should only use the Labby if you are a card counter and can recognize favorable situations. Then you can increase your bet in the Labby fashion at these times, resorting to flat betting one unit at other times. Now, if you are a card counter, or even if you use my chapter 5 method to recognize favorable situations, why do you need a complicated system like the Labby? Just bet a little more at those times a la Min-Max, and you will likely be successful without the cumbersome paper-and-pencil method.

THE "UP AND DOWN," "PYRAMID," "SEESAW," OR "D'ALEMBERT" SYSTEM

Authors have referred to this system by a variety of names. A Frenchman named D'Alembert has been credited with creating it, but the other three names are more descriptive of the method. It goes like this:

1. Start with any initial bet you wish.
2. Decrease your bet one unit after each win.
3. Increase your bet one unit after each loss.

As in any popular system, there is an underlying rationale that captures the interest of the gambler. This particular concept is based on the idea that wins and losses will tend to be equalized, and with this method a profit will be made at equalization time. Consider the hypothetical series:

Hand	Outcome	Next Bet	Profit Status
1	Start	5	0
2	Lose	6	−5
3	Win	5	1 Equalized here
4	Lose	6	−4
5	Lose	7	−10
6	Win	6	−3
7	Win	5	3 Equalized here
8	Lose	6	−2
9	Win	5	4 Equalized here

Notice that the profit goes up every time the number of wins and losses balance out. Moreover, the longer swings (hands four through seven, for example) yield a greater profit. In fact, it is possible on the longer swings to obtain a profit without equalization. A sequence of ten losses followed by eight wins will produce a profit, for example.

With blackjack, you will lose more hands than you win, as we saw previously in the Labby. The extra-money hands (splits, doubles and blackjacks) will make up for this to some degree, as they did for the Labby. An interesting profit trace for the "Up and Down" system is shown in the next graph.

Again, it was necessary to write a special computer program to play through the twenty thousand-hand data package, as punched up on IBM cards. The initial bet was one dollar. A similar run with the initial bet ($5) gave nearly identical results.

While the overall trend is moderately positive for our favorable data package, the oscillations in capital would be intolerable to all but a millionaire. This history shows losing streaks as high as twenty-five thousand dollars and one outstanding winning session of over thirty thousand dollars!

We have already hinted at the reason for these swings. As the game wears on, the bet size grows and grows because of the excess losses. For this sample, though, a five-hundred-dollar bet size was never reached and it was not necessary to start over in defeat. Be advised, however, that the bet size would ultimately grow to five hundred dollars and beyond, if you played long enough. The fluctuations in capital at this point would be horrifying indeed!

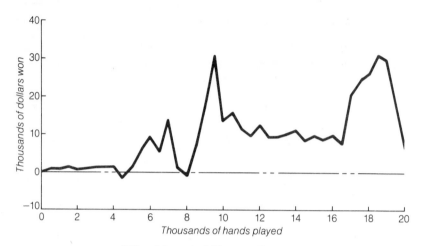

**The Up and Down System
Test Results**

OSCAR'S SYSTEM

This system is treated at great length by Wilson, who became very intrigued when learning of the excellent results that his friend, presumably named Oscar, had with it at craps.[16] Since Wilson did such an excellent job of debunking it with a computer simulation of two hundred and eighty thousand sequences, I refer you to his book for the particulars and will not treat it myself. I only mention it here because it is really a variation of the General Min-Max technique, but which allows you to get in trouble by letting the minimum bet grow. It also has some of the characteristics of a Goodman-type progression. Here are the fairly involved rules for implementing it:

1. Your goal is to achieve a profit of one unit, then restart.
2. Your initial bet is one unit.
3. If you win, increase the bet by one unit unless doing so would cause you to win more than one unit.
4. If you lose, keep your bet equal to the last one you made.

The increasing of your bet after a win is similar to both the General Min-Max and Goodman's progression schemes. The freezing of your bet after losses is also similar to General Min-Max, wherein you always bet the table minimum after losses. The biggest difference, and the killing factor in my opinion, is the freezing of the losing bet values at the *current* level with Oscar's system. This allows a frightening growth in bet size as time wears on, causing the ultimate ruin of the player when the house maximum is finally reached.

Finally, I fail to see any real advantage to restricting yourself to a single unit of profit per sequence. Making profits is what the game is all about. Why place a limit on them? How much better off you would be to limit your *losses* as in the chapter 5 approaches.

"MIMIC THE DEALER" SYSTEM

This system is really a strategy rather than a betting system and has been discussed at length in the literature. All you do here is copy the dealer. You hit until you get 17 or better and never exercise any of the player's options (see "Summing up the Dealer's Rules," chapter 2). This is obviously a disastrous plan. Scarne goes through a calculation and claims that in so doing, you give away an edge of 5.9 per cent to the house.[17] Wilson goes him one better, subtracting the advantage to the player gained by getting an automatic 3 to 2 payoff for blackjacks, worth 2.325 per cent to the player.[18] Scarne's failure to do so, plus other apparent errors, indicate that a more correct value would be a dealer's edge of about 4.5 per cent for this foolhardy approach.

THE "NEVER BUST" OR "PLAY SAFE" SYSTEM

Another very dangerous strategy is to never "bust" your hand. This means standing on all hard totals of 12 or better. While such a method allows you to exercise player's options, it is still a certain losing strategy. The correct plays are shown in chapter 3 for hard totals and are a function of the dealer's "up card." Since no advantage of this information is taken by the "play safe" attitude, such a player's edge will clearly drop well below the practically even game provided by perfect basic play.

NOTES

1. As an aside, you will find some very helpful betting hints for roulette and craps in the section on "General Min-Max" betting.

2. I have also developed a set of equations that predict the distribution of runs over an arbitrary sample size for blackjack. While they produce good agreement with data tested so far, I am not yet ready to publish them. Perhaps I will do so in the next edition of this book, if enough people are interested.

3. While the single zero wheel is standard abroad, as in Monte Carlo, it is fairly scarce in Nevada clubs. Look for it and profit therefrom.

4. Wilson, *The Casino Gambler's Guide* (New York: Harper & Row, 1965), p. 185.

5. Mike Goodman, *How to Win at Cards, Dice, Races, Roulette* (Los Angeles: Holloway House, 1963), pp. 54-60.

6. Ibid., p. 57.

7. Ibid., pp. 51-60.

8. Ibid., p. 54.

9. Ibid., p. 55.

10. John Dana, *Blackjack How to Win the Las Vegas Way* (Las Vegas: Gambling International, 1965).

11. Sidney H. Radner, *The Key to Roulette, Blackjack and One-armed Bandits* (Ottenheimer Publishers, 1963), p. 110-112.

12. Ibid., p. 65.

13. The Count of Monte Carlo, *Blackjack, Vol. 1: A Little Knowledge is a Dangerous Thing* (Ojus, Fla.: President Publications, 1963).

14. Radner, *The Key to Roulette, Blackjack and One Armed Bandits* (Ottenheimer Publishers, 1963), p. 71.

15. Wilson, *The Casino Gambler's Guide* (New York: Harper & Row, 1965), p. 253.

16. Ibid., p. 246.

17. John Scarne, *Scarne's Complete Guide to Gambling* (New York: Simon and Schuster, Inc., 1963).

18. Wilson, *The Casino Gambler's Guide* (New York: Harper & Row, 1965), p. 168.

9. CARD-COUNTING
OR "DECK-CASING" TECHNIQUES

GENERAL COMMENTS

It has never been my aim to discredit those blackjack players who are clever enough to successfully implement card-memory methods. Quite the contrary. If you are sharp enough, and willing to train yourself thoroughly, then you definitely *should* employ one of the techniques discussed in this chapter. They are more accurate and powerful than my chapter 5 methods, to be sure.

Counting methods are not, however, a practical approach for the average person out to have a good time and still retain an edge on the house. They will require intense concentration on your part throughout your bout with the dealer. Moreover, shuffle-up, and the fact that nobody deals to the end of the deck anymore, have eliminated a considerable portion of the advantage card counters can gain.

No attempt at exhaustiveness will be made here, although the elements of the better-known counting techniques will be brought out. If you get really serious about a particular method, you can always beef up your background on this subject by referring to one of the originating authors. These are identified where appropriate.

Card counting is the name used by scientists for what the gambling community often refers to as "deck casing." The latter term is heard often in the casino environment, especially if you are suspected of such a practice yourself. In this case, you may actually be *accused* of being a "case-down" player by a dealer or pit boss as I was on several occasions. That is right. They feel that you are *cheating* the house if you track the cards and they will not hesitate to try to discourage you. See the next chapter for pertinent details, including my discussions with the manager of the former Bonanza Casino in Las Vegas.

The validity of the card-counting concept has been demonstrated time and again with a digital computer. One approach to the problem (Thorp[1]) was to have the computer determine the player's advantage, or disadvantage, when the deck was short-changed by any four cards of a kind and the corresponding "best" strategy was played.

This study showed that the player has an edge if four twos, threes, fours, fives, sixes, sevens, or eights were removed from the fresh deck. A deck that is poor in nines, tens, or aces is advantageous to the house. It turned out that decks without fives, or decks especially rich in ten-value cards, represent the two most clearly advantageous situations for the player. Thus, the prominence of counting techniques based on keeping track of cards with these particular values.

The most outstanding proponent of counting methods is Dr. Edward O. Thorp. His famous book *Beat the Dealer* was a best-seller and his casino exploits are legendary. The first two approaches to be discussed were developed by him and can be pursued in great detail by reading his book.

THE THORP FIVES METHOD

The method of counting fives is highly touted as one of the simplest, winning card-counting systems in existence, which it is. All you do is play the basic strategy, making small bets, until you have seen all the fives dealt out. At this point, you can achieve as much as a 3.6 per cent advantage on the house by increasing your bet and adopting a different playing strategy for the remainder of the deck.

Now, do not expect to get very much "action" out of this system. Rarely will you get to make very many bets after the fives are all out. First of all, this situation normally arises when the deck is just about exhausted, anyway. At a half-filled table, for example, you would get four or five hands out of 100 with this condition, if cards were dealt to the end of the deck. On top of this, the casinos do not deal to the end of the deck normally, and this kills a goodly portion of what would have been bet-raising situations for the fives counter. Finally, the dealers may be on the lookout for the fives themselves and nonchalantly shuffle you out of business every time you get ready to raise your bet. Your realizable percentage with this system is probably well below 0.5 per cent in today's casinos.

Add to this the necessity of learning an additional playing strategy and I think you will find that it is simply not worth bothering with. You are better off with the basic strategy and my Complete System. You will get more action and have a better chance of actually realizing a profit without the necessity for learning another strategy.

To the method's credit, however, is the similarity of the playing strategy sans fives to the basic strategy. The only major differences are seen in the "hard standing numbers." See *Beat the Dealer* by Thorp for strategy details if you are interested.

THE THORP TEN COUNT METHOD

The "ten count" strategy is, to my knowledge, the first widely published version of the powerful card-counting methods. Thorp claims that your edge on the house with this technique will vary from about 1 to 10 per cent.

Such claims are not really valid anymore though, since they are predicated on employing large-scale variations in bet size that are no longer tolerated by the casinos. One casino manager, a very astute individual, assured me that a variation in bet size of 4 to 1 will not be tolerated from a detected card counter since Thorp's book was written. Players varying their bet size at a ratio of more than 2 to 1 may be asked to leave if they are exposed as counters and are winning. Before Thorp, the acceptable bet ratio was as high as 10 to 1 in many casinos. In any event, you are not going to make a fortune counting cards at blackjack anymore, because the casinos are just too alert to this type of player. More about this in the next chapter.

As we have previously discussed, a deck dominated by high cards (tens and aces) is favorable to the player. With the ten count method, you attempt to keep track of the ratio of nontens to tens in the remaining deck. Aces are considered nontens. This ratio (NT/T), adopting Wilson's notation,[2] is a good measure of how rich the deck is in tens. For a full deck the ratio is $36NT/16T = 2.25$. If the ratio drops below this number, then the deck is "ten rich" and is favorable to the player. Should NT/T increase above 2.25, the deck is dominated by low cards and unfavorable.

So, the ten counter makes use of NT/T in two ways:

1. He increases his bet in proportion to the drop in NT/T. He knows, for example, that he has about a 4 per cent edge when NT/T gets close to 1.0 and he plays the basic strategy.

2. He varies his playing strategy in accordance with memorized tables that use NT/T as a parameter. While this is, in general, a real brain stretcher to implement, you can get an idea of its power with an example. If, say, NT/T is very low, then you know the deck is rich in tens. Obviously, you should deviate from the basic strategy in this case, and do less hitting on hard totals of 12–16, thereby considerably enhancing your play. I am not going to dwell on this method because I consider it impractical to implement for most people. Juggling the horrendous array of numbers and strategy variations just requires too much work for what you will ever get out of it. The casinos simply will not let you win anything significant today, especially if they see that you count tens. The method is very well known and watched for in all casinos. It is actually *too* good, if you know what I mean. These comments apply fairly equally to all counting methods. I am sure that before this

book has been around very long, they will also apply to my methods. But at least you need not kill yourself to learn my techniques.

THE SIMPLE POINT COUNT AND HI-LO METHODS

The Hi-Lo technique was presented at the Fall Joint Computer Conference of 1963, in Las Vegas, by a man named Harvey Dubner.[3] With this technique, as with the simple point count method, the cards are classified as follows:

> High cards—10s and aces
> Low cards—2, 3, 4, 5, and 6
> Neutral cards—7, 8, and 9

As a Hi-Lo counter, you must compute the following index before every bet:

$$\frac{\text{Total high cards left} - \text{total low cards left}}{\text{Total cards left}}$$

Try it sometime if you like hard work. At the casino pace it is great fun. Once you have this magic number, you increase or decrease your bet in accordance with a fixed scale. Essentially, the bigger the positive number, the more you bet. If the index is negative, you bet your minimum. No strategy variations are employed, thank heaven.

The simple point count is a little more workable approach and, with some effort, you could train yourself to employ it effectively in a casino. I have used it, but it is not as easy as it appears on the surface. It goes like this:

1. Count plus 1 for every "low" card you see.
2. Count minus 1 for every "high" card you see.
3. Ignore the neutral cards.
4. If the total point count is positive, bet the value of the count.
5. If the count is negative or +1, bet a single unit.

Of all the counting methods, I like this one the best. It is not exceedingly difficult to implement since no mathematical divisions are required and the basic strategy alone is used. Yet it is a fairly precise measure of the richness of high cards in the deck. If you are able to watch and count all the cards in this fashion, you can achieve an edge well in excess of one per cent *without* resorting to strategy variations. You will need to practice for many hours, though, to get good enough to make it worthwhile.

A profit trace of this technique and my own "Complete System" on a ten-thousand-hand sample shows very similar histories. It turns out that the Complete System is a fair *approximation* to this technique and a lot easier to use.

What seems to work out well, with a minimum of strain, is to use my Complete System, but augmented by an occasional rough, simple point count. That is, if the table were clearly covered with an excess of low cards on the last hand, in the simple point count sense, you might take the opportunity to increase your bet slightly. Likewise, decrease it slightly if the last hand obviously used up a lot of high cards. You need not actually bother to get the count, however. You are merely reacting conservatively to obvious gross excesses. This will enhance my Complete System percentages significantly with no real additional work load to the player.

THE THORP "COMPLETE" POINT COUNT

With this technique you keep track of both the simple point count and the total number of unseen cards. Dividing the former by the latter produces another "index" that can be used to vary your bet in accordance with a predetermined scale. The same index is also used to determine strategy improvements. Thorp has still another set of tables for you to memorize and recall instantly, wherein you may accomplish this objective.

Such a method is certainly powerful in theory, but who can really use it? If you are such a person, you will be far better off devoting your effort and unusual mental agility to something worthwhile.

Why waste yourself on a gambling system from which you will probably never be able to profit significantly?

THE THORP "ULTIMATE" POINT COUNT[4]

You can find a description of this method in Wilson although he credits it to Thorp. If you are a real masochist, you can malign yourself beautifully with this "Ultimate" torture.

Here it is:

1. Keep track of tens and nontens.
2. Form the NT/T ratio and use it for strategy variations a la the "ten count" system tables.
3. Keep track of a more complicated point count wherein the card point values range from –9 to +11, and employ this for sizing your bet.

If you think you can do all this at the pace of casino play, why not see if you can get a job replacing a computer. They have been known to rent for as high as $500 an hour. You might even find the work more challenging.

THE WILSON POINT COUNT[5]

Wilson proposes a counting method that is similar to the simple point count, but puts more emphasis on aces. More emphasis, in fact, than can be justified mathematically. Nevertheless, it is a simple and practical approach and his additional suggestions on the counting technique have merit.

He assigns a value of +4 points to all aces, +1 to all ten-value cards and –1 to all others remaining in the deck. A little mental exercise on your part will verify that a complete deck has an aggregate point count of zero, which is tidy and aesthetically attractive.

The in-play counting is implemented as follows:

1. Mentally add 1 for each card dealt.

2. At the conclusion of each hand, subtract 5 points for every exposed ace and 2 points for each ten value card that you see.

This technique is in lieu of subtracting 4 points every time you see an ace, subtracting 1 point as you see tens fall and adding one point for all others. I think it is pretty neat myself, as counting methods go.

You must vary your bet by increasing it in some fashion as the point count becomes increasingly positive. Wilson claims that you can get about a 1 per cent edge with his method using a variation in bets from one to seven units. Unfortunately, the casinos will not let you do this for very long without some sort of harassment. See his *Casino Gambler's Guide* for an extensive discussion of his method. I think you will find it enjoyable.

THE "REVERE" SYSTEMS[6]

There is a book on blackjack by Lawrence Revere and published by a Las Vegas outfit. A number of counting methods are presented with varying degrees of uniqueness. Overall, the book is interesting, colorful, and, I think, more scientifically accurate than most. Moreover, it is to my knowledge the only other book containing Julian Braun's basic strategy for four decks.

Before we launch into our discussions of his systems, though, I cannot resist pointing out a delicious inconsistency. Revere adamantly denies (as do most of the people that I know who are, or have been, connected with a casino) that house cheating is rampant. He scathingly attacks Thorp's claims to the contrary. In particular, he quotes one of my favorite passages from Thorp:

> Now let me ask you again, are men of the Cosa Nostra, who bribe public officials, who steal money off the top, who help to finance their rackets (dope, prostitution, and smuggling) with their casino profits, who commit murder to settle their differences—are these men going to stop short of a little cheating at cards?[7]

Myself, I love this passage and feel that it poses a very provocative question. But Mr. Revere has this to say about it:

This is not a scientific hypothesis. It is by inference alone indicating that members of the Cosa Nostra are everywhere in legalized gambling in Nevada. It infers that because this is true, operators in general will cheat in the same fashion that Cosa Nostra members always cheat.[8]

Revere than goes on to draw an analogy between Thorp's statement and a silly little syllogism that I will not bother to repeat. Please notice, however, that his own statement does not really deny that the Cosa Nostra is, in fact, everywhere in legalized gambling. Nor does it deny that the Cosa Nostra members cheat.

This, however, is not the inconsistency that I mentioned earlier. No more than three pages after this self-righteous plea for scientific accuracy, we find Revere making this statement for scientific posterity:

> I generally play with a girl partner as I am less likely to be spotted as a counter. One girl named Hazel was something special to me. I wanted to win very badly when I played with her. But because I tried so hard, or because of the thought of failure, I could never win in her company.
>
> It is difficult to explain this phenomenon, but it is true. If you do not hope desperately to win but simply play as well as you are able, it seems that you will always win.[9]

I like this passage immensely, both for its early erotic overtones and for the striking scientific objectivity of the closing sentence.

Let us take a look at his counting methods next.

The Revere Five Count Strategy

While this method is very similar to Thorp's version of counting fives, there are some interesting differences. With the Revere version, you also must keep track of the number of cards remaining in the deck, along with the number of fives that are gone. You then vary your bet along a one to four unit scale in accordance with a memorized table parameterizing bet size, total cards remaining, and number of fives gone.

You play the basic strategy when betting one or two units. When the table tells you to wager three or four units however, you play a fives strategy that is almost identical to Thorp's.

Although this version requires more work, it will give you quite a bit more *action*. You will be placing a variety of bets with a higher frequency and investment rate than with the Thorp technique.

The Revere Ten Count Strategy

If you will look at Chart E-1 you will notice that with a count of 3-7, you would be in code five, and you would bet one unit. With this easier strategy there is less room for error on the part of the player[10]

LAWRENCE REVERE, 1969

Despite Revere's claims to the contrary, there is nothing "simple" about his ten count strategy. You count both the tens and the nontens as they fall. If you have seen two tens and five nontens (in the Thorpian sense), then your count is 2-7. Having these two numbers, you then refer to a mentally stored matrix (11 × 17) of numbers containing "codes" and associated bet sizes as a function of your two "count" numbers. Having your code, you then consult three more memorized strategy matrices. The sizes are 11 × 11, 10 × 11 and 14 × 11 for the variable Hit-Stand, Pair Splitting and Double Down strategies, respectively. All this in addition to the requirement for learning the basic strategy!

The primary differences between this and the Thorp ten count are:

1. You must memorize an 11 × 17 "code" matrix instead of computing the NT/T ratio.
2. You count the tens and nontens as they fall, starting with 0 and building up, rather than keeping track of the quantities of these two types that are "remaining" in the deck.

Personally, I could not care less. Only a genius or a maniac would seriously consider such an approach in today's hostile casino environment.

The Revere Plus-Minus Strategy

This is just another variation of the simple point count but is perhaps a bit more accurate. As before, you start with a count of 0, adding a positive number for "low" cards and subtracting a number for "high" cards. The cards are tallied as they fall, as follows:

Card Value	Effect on Count
2 or 7	+1
3, 4, 5, or 6	+2
8, 9	0
10 or ace	–2

Once again you may flex your mind and validate that the total count for a full deck is 0. Just how many variations there could be on this theme staggers the imagination.

But this one has its own peculiar merits. The computer approach of removing four of a kind from the deck did show about a 2 per cent advantage for the player when all sevens are missing from the fresh deck and the optimum strategy is played. Notice that sevens are taken advantage of here, whereas with the simple point count and Hi-Lo method they are not.

Of course, as always with the Revere systems, there is the usual set of additional unwieldy strategy variation matrices to memorize as well. You could make out pretty well with this counting method, however, without the strategy variations. I am sure that it would be fairly effective (say about a 1 per cent overall edge) with a small betting ratio and the basic strategy alone. You might get away with it for a short time, anyway.

The Revere Point Count and Reppert Systems

I have lumped these two together because I do not know very much about them. I do not care to pay anyone for the privilege of teaching me an involved counting method and that is the only way you can obtain these systems. Revere has this to say about the Reppert system.

> There have been many blackjack systems for sale. In most of these, the only winner would be the person who sold the system. One system in particular, "The Reppert System" borders on madness. The values assigned to cards are devoid of reason.[11]

He then goes on to state that the Reppert card values are 48.8 per cent more inaccurate than those in his own plus-minus strategy and 442 per cent more inaccurate than in his own point count. Maybe so, but does it really matter? For most of us, it "borders on madness" to use any of the more involved and *widely published* systems, let alone *paying* to learn about a still more complicated method.

COUNTING TEAMS

A possible approach to alleviating some of the burden on the card counter is to team up with another person. For example, one man could count the cards, compute the necessary ratio and pass it on to his teammate who actually plays the hand. In this manner, one man masters the playing strategy and the other the counting technique. Again, you might get away with this for a short time before the casino operators begin to resent your activities.

A HEAD START

Once in a while you can get a "head start" on an unwary dealer by watching the cards for awhile before you sit down. Stand by a fairly empty table and count the cards until the deck becomes very favorable. This may mean waiting through several shuffles of the deck until a sufficiently juicy situation arises. Now, if the dealer has not noticed you already (and most of them will have done so), sit down and make a "large" bet right away. You have started off with a pretty good edge.

Unhappily, many dealers keep their eye out for this trick and immediately shuffle the deck when you first sit down, wiping away any advantage you might have otherwise gained. In fact, I am sure many dealers *always* shuffle when a new player places his first bet, just to discourage this technique of "striking when the deck is hot" as Thorp terms it.

END PLAY

One of the biggest advantages to the top-notch card counter is termed "end play." If the cards are dealt to the end of the deck, and you are a sharp enough rememberer to realize what the final few cards are, you could periodically sense situations in which you could not lose. For these situations you would bet the table maximum and clean up.

Naturally, the now highly sensitive casinos are aware of this concept, and, consequently, you rarely if ever find a dealer going to the end of the deck. My most recent playing excursions have confirmed that they will now typically shuffle with about fifteen cards left.

COUNTING AND INSURANCE

If you are playing one or another of the card-counting systems, there will be times when it is profitable to take insurance. If, for example, the "ten count" ratio (NT/T) is less than 2.0, it is to your advantage to insure your bet if the situation arises. If the ratio is greater than or equal to 2.0, it is not wise to insure.

It is easy to see the merit to this. As the deck becomes very ten-rich, the chance of a dealer actually having a blackjack is clearly greater.

Obviously, if you knew that *only* tens were left in the deck, then the dealer must have a blackjack if he showed an ace. Naturally, you would jump at the chance to take insurance here.

Another way of looking at the same thing, if you do not care for the NT/T ratio per se, is that more than one-third of the remaining cards must be valued at 10 for insurance to be a good bet. If, for a remote example, you have played the first hand from a fresh deck with the dealer only, and no tens were used, then sixteen of them remain in the deck. Moreover, there are forty-eight cards or less left. So, if the dealer now shows an ace, you would be correct in taking insurance.

For those of you who do not count and therefore prefer one of my systems, you should basically never insure your bet. If, on the other hand, you have definitely seen a great many more nontens than tens depleted from the deck, insurance may prove profitable to you if the opportunity arises on the following hand. A little gross memory combined with common sense can go a long way here.

Finally, I would like to touch upon a subject that has been discussed by numerous authors elsewhere but is nevertheless worthwhile to bring up at this point. That is, despite what many dealers will tell you, it is *not* normally a good idea to insure your bet when you have a blackjack! The fallacious logic they feed you is this: Always insure your own blackjack because, no matter what happens, you cannot lose. If the dealer does have a blackjack, then your regular bet is a push; but you win one unit on the insurance bet (two times one-half of your original one-unit bet). If he does not have a blackjack, you lose half a unit on the insurance bet but get paid at a 3 to 2 rate on your blackjack. In either case you come out one unit ahead!

Okay, this sounds good at first, but we will let Wilson analyze it further for us; he does it so well:

> If you hold a blackjack yourself, and have knowledge of no other cards, insurance is a *very* bad bet. Do not take it. The odds are 34:15 against you for a payoff of only 30:15 (same as 2:1). Your expectation is figured easily as –4/49, or about –8 percent. You are giving the house a big, fat 8 percent when you insure a blackjack![12]

So much for that particular brand of common misinformation. In the next chapter, we will see how the new counting techniques have not only hurt the average player but significantly helped the casinos.

NOTES

1. Thorp, *Beat the Dealer,* p. 48.

2. Wilson, *The Casino Gambler's Guide* (New York: Harper & Row, 1965), p. 116.

3. Ibid., p. 123.

4. Ibid., p. 125.

5. Ibid., p. 105.

6. Lawrence Revere, *Playing Blackjack as a Business* (Las Vegas: Paul Mann Publishing Co., 1969).

7. Thorp, *Beat the Dealer*, p. 142.

8. Revere, *Playing Blackjack as a Business*, p. 16.

9. Ibid., p. 18.

10. Ibid., p. 95.

11. Ibid., p. 19.

12. Wilson, *The Casino Gambler's Guide*, p. 58.

10. THE CASINO BACKLASH

The important thing to remember is that a gambling house is operated for profit. If the method of operation does not assure a profit by fair means, look for the gimmick.[1]

OSWALD JACOBY

COMMENTARY

In my opinion, the gambling casinos of the world are greatly indebted to the mathematicians and computer scientists who have so cleverly devised a myriad of techniques for theoretically beating the game of blackjack. Because of the now widespread knowledge of the existence of these methods, the game is more popular and profitable to the casino than ever before.

Literally millions of tourists (twenty-five million estimated for Nevada alone in 1970) pour into the casinos every year. A very healthy percentage of these people play blackjack. Of some 1,424 registered games[2] in Nevada in 1968, 915 were blackjack tables! There can be no doubt of the game's popularity.

While the number of good card counters has definitely been on the rise, the increase is not sufficient to cause the casinos any alarm. My own observations as a player, and those of many other authors, together with my conversations with the manager of the former

179

Bonanza Casino (he wishes to remain anonymous so I will refer to him henceforth as Mr. M.) have served to identify a number of reasons why the gambling houses no longer fear the card counter.

First and foremost is the fact that a number of legitimate, if incensing, countermeasures have been devised to thwart the effective system player. Secondly, despite the minimization by casino operating personnel of these claims, there is a goodly amount of just plain cheating on the part of the dealers. Whether or not the cheating is fully condoned by the management in all the casinos is debatable. The existence of cheating by blackjack dealers, however, is a hard fact.

Ed Reid and Ovid Demaris, in their truly devastating documentary *The Green Felt Jungle,* describe some of Dr. Thorp's gambling problems.

> In his gambling forays through Lake Tahoe, Reno, and Las Vegas, the professor was victimized by crooked dealers in almost all of the major casinos in Nevada. He has been backed off (thrown out) by pit bosses, he has been harassed by shills, plied endlessly with booze, eyed significantly by plug uglies and, on two occasions, rendered spectacularly rubber-legged and goggle-eyed by knockout drops, courtesy of the house. [3]

While your own experiences will probably be considerably less dramatic, you will be exposed to a variety of house subtleties when you play casino blackjack. A number of these are discussed next.

COUNTERMEASURES

Some of the more effective *legitimate* countermeasures that the casinos employ are:
1. Use of a detective agency to spot known systems players and flag their presence to their casino management clients.
2. Refusal of the very effective player's business at the blackjack table.
3. Use of frequent "shuffle-up" when a counter raises his bet.
4. Complete abandonment of dealing to the end of the deck.

5. Raising the table minimum and lowering the maximum to reduce the possible percentages that can be achieved through a wide variation in bet size.
6. Introduction of dealers who can "count" and shuffle away the favorable situations.
7. Rules variations that are unfavorable to the player.
8. Increased density of the "multiple deck" game.
9. Verbal dealer harassment of the winning player.
10. The use of talkative shills to distract the player.

Our casino manager friend, Mr. M., confirmed what I have long suspected about how the house is warned in advance of an approaching system player. A number of Las Vegas casinos are clients of a private detective agency that specializes in helping them spot card counters or, as in my case, someone who they *think* is counting.

Now, on the first weekend that Phyllis and I tested out my "Complete System" in Las Vegas, we both won at a truly alarming clip for our puny bet sizes. By the time our combined hand total approached 1,600, we had piled up a win of some $200. This may not sound like much to you if you have never played in a casino, but it really is outstanding for our bet variation.

We would play for hours on end and break even, or lose a few dollars. The fact of our longevity on a small investment was, in itself, enough to arouse the interest and resultant displeasure of most of the dealers we played against. When things really started to go our way, though, and we would creep steadily ahead, the pit boss would invariably study our play with raised eyebrows.

We made absolutely no attempt to disguise ourselves as we moved about downtown. In some places we *felt* that we were cheated, in others we definitely were not, and in still others, we are absolutely certain that the dealers deliberately and specifically cheated us. I will tell you about some interesting and humorous instances of this later. More to the point, since we were not camouflaged in any way, it was easy for detectives to spot us and forewarn the casinos.

As the weekend progressed, the house harassment increased. As we passed through the lavish entrance to one of the newer sin palaces on the Strip, Phyllis noticed two men standing near the door. They were pointing directly at my back and gesticulating wildly to someone in the pit area.

We approached a single-deck table and watched. Immediately, a sharp new dealer took over and started chattering to a new player (probably a shill) about how some people "fight" the game. Just to see what would happen, I sat down and started to play.

As a test, I kept track of a rough simple point count. Sure enough, he shuffled away all favorable situations as fast as they came up. Consequently, my bet size remained constant. Becoming annoyed, but now curious to see what he would do, I deliberately tripled my bet for no good reason. This stopped him cold for a good half a minute. He just stared at the increased bet in disbelief. "Could I have miscounted?" he must have wondered. "If he was such a threat that they had to call upon my talented services, why did he make such a stupid bet?" The situation was ironic, for I happened to win that hand. With a self-satisfied smile, and just as if I had planned the whole thing, I got up and walked out of the casino.

Mr. M. recounts a tale that demonstrates the respect that he and other casino managers have for well-known counting experts. One day as Reppert (who heavily advertises and sells a counting system of his own) walked into the Bonanza, one of his spotters tapped Mr. M. on the shoulder and whispered, "Hey, there's Reppert." Mr. M. walked right over to him, smiled and said, "Hello, Mr. Reppert." Reppert, with furrowed brow, answered quizzically, "Do I know you?" Mr. M. responded, in dead earnest, "No, but I know you and I don't want any of your twenty-one business at my tables." Reppert, somewhat taken aback, blurted, "Gee, that's the first time I've ever been told to leave before I even started to play." Maybe so, but I bet it will not be the last.

Among other things, Mr. M. admitted to the following as we sat in his paneled office containing two multiselector closed circuit TV monitors and one of the most complete libraries on casino gambling that I have ever seen:

1. Cheating by some dealers does exist, although he feels that the claims of Thorp and others are magnified "at least 100 times" over actual facts.
2. Bet variations of greater than 4 to 1 will no longer be tolerated from suspected counters. A 2 to 1 variation is fairly safe.

3. Dealers do occasionally harass systems players to discourage them.

4. What I told him about my Complete System was valid, and, in fact, he had come up with some elements of it himself. (He, in essence, had also discovered the seven-card concept, although he did not have the accurate percentages that I have. He also said that it would be profitable to decrease your bet after a blackjack. This is a special case of my four-card rule.)

5. The "folding your hand" option (see chapter 2) that is offered in his casino is generally favorable to the house. It can be advantageous to the card counter but is actually used as just another means for spotting them.

6. He would kick out a winner using the Du-Rite Wheel. He said that he would even ask a simple flat-betting basic-strategy player to leave if he were taking up valuable space when the house was crowded. Most casinos, however, will not do these things.

7. He has studied horses as well as blackjack and has come to some conclusions of his own. After considerable thrusting and parrying, I was able to determine that his research had been along lines similar to those of mine, and that we have come to some common conclusions. Going me one better, he claims to have already implemented a system at a track that enables him to win $100 to $200 a day. For a man like him, this was not enough to warrant his daily presence at the track and he stopped messing with it. I found this so encouraging that I cannot wait to finish this book and expand on my own effort. You cannot be cheated or shuffled against, as an individual anyway, at the track!

8. Finally, when asked how he could readily tell a counter from a cheater (aces-up-the-sleeve types, you know), he smiled and said, "Cheaters are usually rather scruffy-looking, while counters are generally very academic in appearance." So, in the casino environment at least, the characters tend to live up to their stereotypes. If you happen to be unfortunate enough to somehow appear both "scruffy" and "academic" simultaneously, you do not have a chance.

CHEATING

Now, as promised, some observations on cheating. There are some dealers who apparently can give you any card that they wish whenever they please. I received an astonishing exhibition of this, or took part in an even more astonishing coincidence, on a stopover in Las Vegas.

The Apollo 11 simulation team for the U.S.N.S. *Redstone* and U.S.N.S *Huntsville* tracking ships (at the time, I was team chief and computer advisor for the Department of Defense) was on the way home from Hawaii and we made a one-night rest stop at Las Vegas before continuing on to Goddard Space Flight Center and Cape Kennedy. I went downtown to my favorite casino and looked over the fairly crowded tables.

One table, sporting a two-deck game, particularly attracted my attention because it was empty. I sat down, kind of uneasily, wondering why this table was empty, and placed a one-dollar bet. On the very first hand, the smirking dealer gave himself a blackjack. I looked at him in disgust, suspecting the worst, and hesitated before placing another bet. As I started to get up and leave, he said, "Wait a minute, what's the matter?" I answered, "What's the matter? Are you all right? You just gave yourself that blackjack. No wonder your table is empty." He looked hurt and said, "Play. It's all right."

Still debating with myself, I placed another bet and received a pair of sixes with the dealer showing both a five and a self-satisfied smile. Splitting them, I received another six followed by an additional three sixes! I dutifully hit all six hands, after splitting, and received a high card on each. The dealer was obviously pleased with himself, as I stood pat now and waited for him to cream me. But no! He was really beside himself with pride in his own ability. He called the pit boss over, for the umpteenth time no doubt, and said, "How about that!" The pit boss gave him a sort of bored or disgusted look, I am not sure which, shrugged his shoulders and walked away in silence.

Getting tired of all this by now, I said angrily, "Let's get this over with. Beat me and be done with it." With a very hurt look, the dealer turned over a ten, giving himself a hard 15. He then looked me in the eye and busted with a picture card.

In disbelief, I collected my winnings and placed another bet. I guess he had run out of tricks, because I won that hand, too. The dealer then transfixed me with one of those "you and I share such intimate secrets" looks and said bold-facedly, "You won that hand fair and square, understand?" I understood all right. Answering, "Roger that!" I grabbed my chips, cashed in and got the hell out of that clip joint like Apollo 11 on its way to the moon. In retrospect, I wish that I had stayed on to see what he would do next, don't you?

While some dealers are swift indeed, others often fail to make the grade. During one of the casino testing weekends for the Complete System, I wandered into a small downtown casino, sat down, and proceeded to win forty-five dollars in a matter of minutes. To make matters worse, another small bettor at the same table did almost as well.

The annoyed pit boss replaced the hapless dealer with a young kid who was obviously very nervous. It must have been his first cheating assignment, or something. Initially, he seemed to have some facility with the cards, but his shuffling looked very strange. The other player and I began to stare at him, fascinated with the unusually long shuffle and his wild hand motions, which were growing more uncertain and erratic by the moment. Suddenly, he lost control completely. The *cards* were beginning to run away with *him*! With a terrified look in his eyes, and under the alarmed gaze of the pit boss, he gave the deck a sweaty-handed squeeze and all fifty-two cards ejaculated from his hand. They followed a short, arcing trajectory, spreading out as they cascaded down about my head and ears. Choking with laughter, the other player and I got up and left the stricken dealer, exasperated pit boss and littered table.

There is an infinite variety of ways in which a dealer can cheat you if he has the inclination. While I do not wish to overemphasize the sordid aspects of the game, some of the more common cheating methods should be mentioned.

Deck Stacking

By deck stacking, I am referring to the arrangement of the cards in some special order as the dealer picks them up. He must then preserve this arrangement with a phony cut and shuffle. If successful, he then clobbers you with a preset or "stacked" deck.

The Hi-Lo pickup is one of the most common forms of deck stacking. The dealer alternately picks up a high and a low card in a continuing sequence as he gathers in the spent cards. If he manages to preserve this order, then you and everyone else will get a lousy hand. You might get a ten, six, the player next to you a nine, five, and so on down the line.

This is really not too difficult to spot if you are on the lookout for it. Normally, the dealer will simply scoop up all the cards on the table in one swift motion, without paying any particular attention to the individual cards. If you see a dealer picking up the cards in little jerky motions, something is probably amiss. Watch his eyes. If he is staring intently at the cards as he rapidly works them over in the pickup, he is probably stacking.

How effective his stacking is will depend on his skill. He has to work fast and will occasionally make errors. That is why he does not beat you every time, even with a Hi-Lo deck.

Anyway, I have not seen quite as much of this lately as I did a few years back. At that time, it was possible to wander around downtown Las Vegas and spot suspicious-looking pickups all over the place. It was unbelievable. Invariably those dealers with rapid, but erratic, pickup motions would kill off a table in short order.

On my last few times out, however, there was very little of this type of cheating in evidence. It seems to have been replaced by a simple new trick, which all dealers can perform. That is, the dealer gives himself blackjack on the first hand after a shuffle. This is so easy to pull off, and so effective, that it seems to have taken the number-one spot in the hearts of the cheat dealers. Almost any dealer can save himself an ace and picture on the bottom or top of the deck. With a fairly full table, the deck is shuffled once every couple of hands or so. By this simple tactic, even a sweet, innocent young female dealer in Reno can guarantee herself a win rate as high as 50 per cent if she wishes.

I have played against one dealer in Reno who would get blackjack on the first hand after a shuffle at least once out of every three times. Now, the probability of the dealer getting a blackjack from a fresh deck is $2 \times 4/52 \times 16/51 = 0.048$, or about one hand out of twenty-one! Just remember this whenever you play in Nevada.

A still easier way for the dealer to gain an advantage by a form of deck stacking is merely to make sure that he gets an ace after every shuffle. This way, he need only monkey around with one card. Yet, look at the advantage he has with that guaranteed ace every two or three hands. It means he will get a blackjack 31.4 per cent of the time after every shuffle. His probability of having at least a two-card 19 is now 24/51 or 47 per cent! Suffice it to say that no system can overcome this type of cheating. If you find a dealer getting an ace after all or most of his shuffles, change dealers, or casinos. After all, this situation should only average some two hands out of thirteen.

More complicated stacking techniques such as the "Kentucky step-up"[4] are probably also used, but I have never discovered it. This is just a clump of cards preserved in the sequence 7, 8, 9, 10, 10, J, Q, K, A. The dealer fixes it so that you cut the seven to the top of the deck; or else he manipulates it to this position by sleight-of-hand. The seven is burned and you get eight, ten and the dealer nine, ten followed by your jack, king to the dealer's queen, ace.

Actually this sequence need not be applied too rigorously to be just as deadly. Any card could replace the seven, for example, since it is burned anyway. Any five ten-valued cards, and in any order, could be placed between the nine and the ace also. So, due to its numerous possible variations, it is not quite so easy to spot. A clever reader could no doubt invent all sorts of murderous little clumps such as these. Imagine what nightmares a professional cheat dealer can concoct!

Implicit in most of these techniques are phony shuffles and cuts. These are discussed next.

False Cuts and Shuffles

Here are two of the more common methods of faking a cut:
1. *The crimp.* The divided pack is given a crimp by slightly bending the upper half of the deck upward and the lower half

downward. This leaves a small bowed space between the deck halves. Although the unsuspecting player will cut the deck precisely at the crimp most of the time, a miscut is easily rectified by the card sharp. He can easily spot or feel the crimp and unobtrusively make the necessary adjustment of a few cards fairly easily. The crimp technique has been demonstrated to me by a real pro and it really works. He could make me cut the deck anywhere he wanted with that crimp.

2. *The pickup switch.* Here, the bottom half of the divided pack is picked up with one hand, as if to place it on top of the others. Instead, the original top half is picked up with the same hand after transferring the original bottom half to the other hand. The original top is now placed back on top. This technique works best when the dealer has been varying his pickup method all along. He sometimes places the bottom on the top half before picking up the whole deck. Other times, he places the top half in one hand and covers it with the original bottom cards. Then, without warning, he pulls the pickup switch. It can be easily overlooked by the unwary player.

There are all sorts of fake shuffles. Radner[5] describes the "pull through" shuffle, the "running up" of hands with the "Haymow" shuffle and more. His book is definitely recommended reading if you are really out to spot the cheating. However, there is nothing like a demonstration of some of these fantastic maneuvers to really convince you of their power. It would be worth your while to try and locate such a demonstration in your area before you take on the pros in Nevada. Some of those guys must send their decks to obedience school.

A related, and extremely common, form of cheating is the dealing of "seconds."

Dealing "Seconds" or "Second Carding"

Off and on during the test of the Min-Max System in Las Vegas, I must have played a total of four or five hours with a pleasant gray-haired dealer in a large downtown casino. His relaxed attitude and obvious lack of concern for my small, but long-term, winnings drew me to his table time and again.

Just before we left for Los Angeles late Sunday afternoon, I decided to pay him one final visit. Right away I noticed a complete change in his attitude and technique. Now, he was a very quick, red-faced and exceedingly businesslike professional. It was as if he had somehow shifted gears or stepped back into his own past as a hustler. I do not know; that particular casino must have finally gotten sick of my small but constant drain on their resources and given him the green light to do me in.

I just could not seem to put together a winning sequence. He really had the edge on me and I could not spot the problem. Finally, he gave himself away. The table was nearly full and I was two seats away from first base. This meant that I would receive the third card dealt. As I watched him throw out the cards, they went in sequence: all the way across the table, all the way across the table, halfway across the table. When he dealt out the next hand, this same funny sequence was repeated. I could not miss it because I had to reach way out for my cards, but nobody else did.

He was dealing me "seconds." This is a form of cheating in which the dealer peeks at your card (often by squeezing the edges together and checking the space under the hump for either a picture or lack of pips) and, not liking it, passes you the next or "second" card.

The problem he was having is a common one for second carders. The action of the upper and lower cards on the second will sometimes retard its progress because of friction. Unless the dealer uses a strong flick of the wrist, that card will not follow the same trajectory as a normally dealt one. Rather, it will fall a little short, as his did. If your cards start to nosedive like this, it is time to vamoose.

Other Methods

Many other forms of cheating are practiced, I am sure. For example, card inking, scratching, nicking and removal from the deck may occasionally be employed. These techniques, however, are necessarily exercised with extreme caution because of the tangible evidence they produce. The most common and dangerous horn-swoggling methods, therefore, are in a class with those previously discussed.

WHAT ABOUT THE LITTLE GUY?

Many of the books and articles that I read in studying blackjack made frequent references to house cheating, particularly in Las Vegas. These documents, as well as my own experiences, bear out the fact that some casinos will not stop at cheating the little guy, the one- and two-dollar bettor, if they feel that they would otherwise be beaten consistently by him.

A piece of fallacious reasoning that is commonly offered by many self-styled gambling experts goes something like this. Always try to play at a table where at least one player is betting very high. The dealer, they reason, will be so busy contending with the high bettor that he will not even bother to consider your paltry one or two dollars. Bull! The tables with the big bettors are precisely those that are likely to be blessed by a cheating dealer. If he employs a technique such as deck stacking, he will get you right along with the big spenders. Not only that, why not cheat you as well while he is at it? If you are going to cheat, cheat. Why go only halfway?

Also, if you can win consistently at a small bet size, you can probably do even better at higher stakes, as we saw from chapter 6. So, the casinos are not about to let you win so much at small bets that you can afford to make larger ones with comfort. Professional casino operators are also cognizant of the concepts explored in chapter 6.

Being a "little guy," then, is not necessarily a protective buffer against the wrath of the wounded casinos.

SUMMATION

With its arsenal of legitimate countermeasures, backed up by dealers who can and will cheat if necessary, the casino is no longer in any real danger of being "taken" by a talented blackjack system player. Thanks to the scientific community, so much good information on counting systems has now been published that the casinos are ready and waiting to pounce on anyone who wins consistently.

Because of this "backlash," I sincerely feel that it is senseless for the blackjack enthusiast to waste a great many hours learning complicated counting strategies.

Thanks to this book, it has been demonstrated for the first time that there really is "no need to count." You can have the pleasure of fighting the backlash without the pain. My formula for this is:

1. Learn the "basic" or reduced strategy.
2. Carry your Du-Rite Wheel for quick checks on difficult-to-remember plays.
3. Use either the Min-Max or Complete System approach to betting and "lean" whenever you can.
4. Bet small and keep your variation at 4 to 1 or less.
5. Move around. Change tables and casinos frequently.
6. Keep a sharp eye out for possible dealer cheating along the aforementioned lines.

If you follow this advice, you will have a nice little edge on the house. You should win more often than you lose. When you do lose, it will be precious little—and only after a long battle. When you win, it will be occasionally significant. With a good run of cards, you can often pick up a hundred or two on a weekend and have a ball. If you lose a little now and then in Las Vegas—well, you still had a great time. That is my philosophy. For those of you who can honestly afford to implement it, have a go at it and good luck!

NOTES

1. Oswald Jacoby, *How to Figure the Odds* (Garden City, N.Y.: Doubleday & Co., 1947), p. 52.

2. Revere, *Playing Blackjack as a Business*, p. 141.

3. Ed Reid and Ovid Demaris, *The Green Felt Jungle* (New York: Pocket Books, Inc., 1964), p. 207.

4. Wilson, *The Casino Gambler's Guide*, p. 149.

5. Sidney H. Radner, *How to Spot Card Sharps and Their Methods* (New York: Key Publishing Co., 1957), pp. 49–61.

A. GAMBLER'S RUIN FORMULAS

The *Gambler's Ruin Problem* has been treated extensively by mathematicians.[1] It may be stated as follows in classical terms:

Given: p = probability of winning 1 unit
q = probability of losing 1 unit
S = gambler's starting capital
C = combined capital of gambler and adversary

Find: R_S = probability of gambler's ultimate ruin

For an "unfair" game ($p<q$), we should also like to find the expected duration of the game (D_S) to complete the classical problem. However, for our systems we have ($p>q$), and the expected game duration is infinite. That is, if you have the edge over the casino, the game may go on forever.

On the theory that you will pursue the mathematics in the referenced texts if interested, I will give only a cursory development here. A clear presentation of the formulas is my goal, so that you may apply them to your needs in cookbook fashion.

Letting:

R_{S+1} = chance of ruin with one additional unit of capital

R_{S-1} = chance of ruin with one less unit of capital

and combining probabilities produces a difference equation of the form:

$$R_S = p^R{}_{S+1} + q^R{}_{S-1}. \tag{1}$$

A solution to (1) has the form:

$$R_S = A + B \cdot (q/p)^S. \tag{2}$$

Now, if the player's starting capital (S) is wiped out, then his probability of ruin is 1.0. If he accumulates all of the combined capital (C), he has broken the bank of his adversary and his ruin probability has vanished. These concepts may be expressed:

$$R_{(S-0)} = 1$$
$$R_{(S-C)} = 0. \tag{3}$$

Imposing the boundary conditions (3) on the relationship (2), solving the resulting two equations simultaneously for the constants A and B, and combining[2] yields the desired formula:

$$R_S = \frac{(q/p)^C - (q/p)^S}{(q/p)^C - 1} \tag{4}$$

Equation 4 is the basic ruin formula. Of course, the casino is, in effect, infinitely rich and you have no hope of cleaning out, say, a Caesar's Palace or a Harrah's. But you may use (4) in a different manner as I have done in computing the curves in chapter 6. In this case let:

C = amount to which you desire to increase your capital

that is, net gain = $C - S = G$ $\qquad$ (5)

and q, p, and S are as previously defined. Now we have:

R_S = probability of ruin prior to gaining G units

and:

$1 - R_S$ = probability of gaining G units successfully.

To obtain the "ultimate" ruin probability as plotted in chapter 6, let C be infinite in equation 4. Thus

$$(q/p)^C \to 0$$

since $q < p$ for a favorable game, equation 4 reduces to:

$$R_S = (q/p)^S \tag{6}$$

NOTES

1. William Feller, *An Introduction to Probability Theory and Its Applications* (New York: John Wiley & Sons, Inc., 1957), pp. 313–318; Michel Loeve, *Probability Theory* (New York: D. Van Nostrand Co. Inc., 1960), p. 48; Emanuel Parzen, *Modern Probability Theory and Its Applications* (New York: John Wiley & Sons, Inc., 1960), pp. 144–145; Frank Spitzer, *Principles of Random Walk* (New York: D. Van Nostrand Co. Inc., 1964), pp. 217–218; Wilson, *The Casino Gambler's Guide*, pp. 287–289.

2. The algebra here is really trivial once you have accepted equation 2. Go ahead and give it a shot and *satisfy* yourself.

B. DERIVATION OF THE BASIC PROFIT EQUATIONS

Let the data be collected and grouped in runs. Let the total number of winning runs of length i (where $i = 1, 2, 3 \ldots \ldots n$) be given the symbol a_i and the total number of losing runs of length i be denoted b_i. Schematically, the runs and the corresponding symbols representing the *total number* of each kind, are shown below:

b_1	b_2	b_3	b_4	*Bet ($)*	a_1	a_2	a_3	a_4	*Bet ($)*
L	L	L	L	X_0	W	W	W	W	Y_0
	L	L	L	X_1		W	W	W	Y_1
		L	L	X_2			W	W	Y_2
			L	X_3				W	Y_3
				etc.					etc.

As shown in the above diagram, the amounts that were bet on the first loss or win of a run are *assumed* to be X_0 and Y_0, respectively. Similarly, the second bet in a series is X_1 or Y_1, the third is X_2, Y_2, etc.

Note that the actual *amount* that was bet and lost on the *first loss* in each losing run is a variable. The amount of this bet will depend upon which type of winning run of Ws it is closing out. For example, if a first loss closed out a winning run of length 3, then the amount

195

bet on that particular loss would correspond to Y_3. This example is next illustrated:

Outcome	Bet ($)
W	Y_0
W	Y_1
W	Y_2
L	$Y_3 = X_0$

Similarly, the amount of money won on a *first win* of a winning run is also a variable. Its value will depend on what type of a losing run it is closing out. If, for example, a win closes out a run of five losses, the amount of money won on that bet will correspond to the value X_5. That is:

Outcome	Bet ($)
L	X_0
L	X_1
L	X_2
L	X_3
L	X_4
W	$X_5 = Y_0$

In actual practice, therefore, nominal values of X_0 and Y_0 will never be bet. Rather, these first bets of a run will *always* correspond in value to the close-out bet of the preceding run.[1]

We are now in a position to write down expressions for the total amount of money lost (L_0) and won (W_0) exclusive of factors such as blackjacks, split pairs, and hands in which we double down.

Now, the number of first losses on which Y_1 dollars were bet equals the number of winning runs of length one. The number of winning runs of one in a row, by definition, equals a_1. Thus the total money lost on a Y_1 bet $= a_1 Y_1$. Likewise, the number of first loss bets of Y_2 in value $= a_2$ and the total dollar value for this equals $a_2 Y_2$.

Following this line of reasoning, we may write:

$$L_1 = \text{Total money lost on first losses} = a_1 Y_1 + a_2 Y_2 \quad (1)$$
$$+ \dots a_n Y_n.$$

In like fashion, we may deduce:

$$W_1 = \text{Total money won on } \textit{first} \text{ wins} = b_1 X_1 + b_2 X_2 \quad (2)$$
$$+ \dots b_n X_n.$$

To determine the total money lost on *second* losses we must multiply the amount of money bet (which is always $= X_1$) by the total number of bets of this type made $(b_2 + b_3 + b_4 \dots b_n)$. Proceed in the same fashion to obtain money lost on *third* losses and *fourth* losses etc. This gives:

Total lost on:

$$\text{Second Losses} = (b_2 + b_3 + b_4 + \dots + b_n)X_1 = X_1 \sum_{i=2}^{n} b_i$$

$$\text{Third Losses} = (b_3 + b_4 + b_5 + \dots + b_n)X_2 = X_2 \sum_{i=3}^{n} b_i$$

$$\text{Fourth Losses} = (b_4 + b_5 + b_6 + \dots + b_n)X_3 = X_3 \sum_{i=4}^{n} b_i \quad (3)$$

.

.

.

.

.

.

$$n^{\text{th}} \text{ Losses} = b_n X_{n-1}.$$

The total amount of money lost (L_0) may finally be obtained from:

$$L_0 = L_1 + X_1 \sum_{i=2}^{n} b_i + X_2 \sum_{i=3}^{n} b_i + X_3 \sum_{i=4}^{n} b_i + \dots + X_{n-1} b_n. \quad (4)$$

The total amount of money won may, of course, be obtained by identical reasoning. The application of such gives:

$$W_0 = W_1 + Y_1 \sum_{i=2}^{n} a_i + Y_2 \sum_{i=3}^{n} a_i + Y_3 \sum_{i=4}^{n} a_i + \dots + Y_{n-1} a_n. \quad (5)$$

Now, we will note that any term, such as (L_1), in the loss equation may be treated as a negative win. Conversely, any term, such as (W_1), in the win equation may be considered as a *negative loss*. We can thus remove the quantity (L_1) from equation 4 and add it to equation 5 as $(-L_1)$. Also, we can remove the quantity (W_1) from equation 5 and place it in equation 4 as $(-W_1)$. This gives the following two equations.

$$L_0 = X_1 \sum_{i=2}^{n} b_i + X_2 \sum_{i=3}^{n} b_i + X_3 \sum_{i=4}^{n} b_i + \ldots + X_{n-1}b_n - W_1 \quad (7)$$

$$W_0 = Y_1 \sum_{i=2}^{n} a_i + Y_2 \sum_{i=3}^{n} a_i + Y_3 \sum_{i=4}^{n} a_i + \ldots + Y_{n-1}a_n - L_1 \quad (8)$$

Substituting the values of W_1 and L_1 from equations 1 and 2 into equations 7 and 8 and combining terms, yields:

$$L_1 = (\sum_{i=2}^{n} b_i - b_1)X_1 + (\sum_{i=3}^{n} b_i - b_2)X_2 + \ldots + \quad (9)$$
$$(b_n - b_{n-1})X_{n-1} - b_n X_n$$

$$W_0 = (\sum_{i=2}^{n} a_i - a_1)Y_1 + (\sum_{i=3}^{n} a_i - a_2)Y_2 + \ldots + \quad (10)$$
$$(a_n - a_{n-1})Y_{n-1} - a_n Y_n$$

Finally, we can write these equations in a highly enlightening form. That is:

$$L_0 = B_1 X_1 + B_2 X_2 + \ldots B_n X_n \quad (11)$$
$$W_0 = A_1 Y_1 + A_2 Y_2 + \ldots A_n Y_n \quad (12)$$

Where the As and Bs are constants determined from the statistics (as and bs) that represent the actual distribution of winning and losing runs. The profit (P_0) is simply:

$$P_0 = W_0 - L_0 \quad (13)$$

The formulas for the As and Bs have been obtained directly from equations 9 and 10 and are:

$$A_0 = \sum_{i=1}^{n} a_i \qquad\qquad B_0 = \sum_{i=1}^{n} b_i$$
$$A_1 = A_0 - 2a_1 \qquad\qquad B_1 = B_0 - 2b_1$$
$$A_2 = A_1 + a_1 - 2a_2 \qquad B_2 = B_1 + b_1 - 2b_2$$
$$A_3 = A_2 + a_2 - 2a_3 \qquad B_3 = B_2 + b_2 - 2b_3 \quad (14)$$
$$A_4 = A_3 + a_3 - 2a_4 \qquad B_4 = B_3 + b_3 - 2b_4$$
$$\vdots \qquad\qquad\qquad \vdots$$
$$A_{n-1} = A_{n-2} + a_{n-2} - 2a_{n-1} \qquad B_{n-1} = B_{n-2} + b_{n-2} - 2b_{n-1}$$
$$A_n = -a_n \qquad\qquad B_n = -b_n$$

NOTE

1. The only exception to this rule will be on the first bet of the entire series. For this bet, and only this bet, no series is being closed out and the true value of Y_0 was used. If this first bet is a win then we will define two statistics as follows:

$$a_0 = 1$$
$$b_0 = 0$$

If this first bet is a loss then, by definition,

$$a_0 = 0$$
$$b_0 = 1$$

The profit equations 11 and 12 would then become

$$L_0 = B_1 X_1 + B_2 X_2 + \ldots + B_n X_n + b_0 X_0 \tag{11a}$$
$$W_0 = A_1 Y_1 + A_2 Y_2 + \ldots + A_n Y_n + a_0 Y_0 \tag{12a}$$

Of course, for the case of a great many hands, the first hand of the entire series may be ignored, thereby eliminating the necessity for considering the statistics a_0 and b_0.

C. INCORPORATION OF THE "EXTRA MONEY" TERMS

As described in chapter 7, let us assume that we have collected the following set of "extra money" statistics:

Y_iWS = Total number of "win splits" on a Y_i bet.

Y_iLS = Total number of "lose splits" on a Y_i bet.

X_iWS = Total number of "win splits" on an X_i bet.

X_iLS = Total number of "lose splits" on an X_i bet.

Y_iWD = Total number of "win doubles" on a Y_i bet.

Y_iLD = Total number of "lose doubles" on a Y_i bet.

X_iWD = Total number of "win doubles" on an X_i bet.

X_iLD = Total number of "lose doubles" on an X_i bet.

Y_iB = Total number of blackjacks on a Y_i bet.

X_iB = Total number of blackjacks on an X_i bet.

Where $i = 1, 2, 3, \ldots \ldots n$.

Now, the first term in our basic win equation, as derived in Appendix B, is $A_1 Y_1$. When this term is corrected to include the "extra money" it will be denoted $A_{1T} Y_1$ where:

$$A_{1T} = A_1 + \Delta A_1$$

Similarly:

$$A_{2T} = A_2 + \Delta A_2$$
$$A_{3T} = A_3 + \Delta A_3$$
$$\cdot$$
$$\cdot$$
$$\cdot$$
$$\cdot$$
$$A_{nT} = A_n + \Delta A_n$$

Our problem at hand is to find expressions for the ΔA_i in terms of extra money statistics that we have collected. First, let us find an expression for ΔA_1. The total number of win splits that occurred on a Y_1 bet is denoted Y_1WS. Since this is a total of wins it must be added to A_1. The same thing applies to Y_1WD.

Since blackjacks only result in half an extra win (pays 3 to 2), we must divide the total number that occurred on Y_1 bet by two and add to A_1.

The total losing splits on a Y_1 bet (Y_1LS) and losing doubles (Y_1LD) must be subtracted from A_1.

We can now write out the ΔA_i as:

$$\Delta A_1 = Y_1\mathrm{WD} + Y_1\mathrm{WS} + \tfrac{1}{2}Y_1\mathrm{B} - Y_1\mathrm{LD} - Y_1\mathrm{LS}$$
$$\Delta A_2 = Y_2\mathrm{WD} + Y_2\mathrm{WS} + \tfrac{1}{2}Y_2\mathrm{B} - Y_2\mathrm{LD} - Y_2\mathrm{LS}$$
$$\Delta A_3 = Y_3\mathrm{WD} + Y_3\mathrm{WS} + \tfrac{1}{2}Y_3\mathrm{B} - Y_3\mathrm{LD} - Y_3\mathrm{LS}$$
$$\cdot$$
$$\cdot$$
$$\cdot$$
$$\cdot$$
$$\Delta A_n = Y_n\mathrm{WD} + Y_n\mathrm{WS} + \tfrac{1}{2}Y_n\mathrm{B} - Y_n\mathrm{LD} - Y_n\mathrm{LS}$$

In the same manner, let us find a set of corrections (ΔB_i) for extra money to the basic loss equation. Here we are interested in the extra money statistics that pertain to the X bets. The equations for the ΔB_i take the same form as the ΔA_i equations. Thus:

$$\Delta B_1 = X_1\mathrm{WD} + X_1\mathrm{WS} + \tfrac{1}{2}X_1\mathrm{B} - X_1\mathrm{LD} - X_1\mathrm{LS}$$
$$\Delta B_2 = X_2\mathrm{WD} + X_2\mathrm{WS} + \tfrac{1}{2}X_2\mathrm{B} - X_2\mathrm{LD} - X_2\mathrm{LS}$$

$$\Delta B_3 = X_3 WD + X_3 WS + \tfrac{1}{2}X_3 B - X_3 LD - X_3 LS$$

.

.

.

.

.

$$\Delta B_n = X_n WD + X_n WS + \tfrac{1}{2}X_n B - X_n LD - X_n LS$$

However, as written, these corrections represent wins, therefore they must be treated as negative losses in the loss equation. We get:

$$B_{1_T} = B_1 - \Delta B_1$$
$$B_{2_T} = B_2 - \Delta B_2$$

.

.

.

.

.

$$B_{n_T} = B_n - \Delta B_n$$

The actual win, loss and profit equations are:

$$W_T = A_{1_T}Y_1 + A_{2_T}Y_2 + A_{3_T}Y_3 + \ldots \ldots + A_{n_T}Y_n$$
$$L_T = B_{1_T}X_1 + B_{2_T}X_2 + B_{3_T}X_3 + \ldots \ldots + B_{n_T}X_n$$
$$P_T = W_T - L_T$$

D. THE DU-RITE STRATEGY WHEEL

On the next page, you will find a reproduction of a handy little device conceived by my wife, Phyllis. The complete basic strategy is contained in the circular design on its face. The device enables you to dial the correct play instantly for any combination of your hand total and the dealer's up card.

A beautiful vinyl Du-Rite Wheel may be obtained for four dollars (price subject to change without notice) by writing to the following address:

Du-Rite Company
9 Abbott Street
West Babylon, Long Island
New York 11704

The attractive open-faced layout of the Du-Rite Wheel makes an excellent learning aid for mastering accurate blackjack, and, at this writing, may still be employed in most casinos. They come in very handy when you get a little flustered by a large bet and wish to check yourself. I have used them in many Nevada casinos and in no case did the casinos frown upon them.

Use of the Wheel

The Du-Rite at 21 Strategy Wheel

The general procedure to obtain a correct play is merely:
1. Turn the pointer until your hand total, or pair, is centered at the top of the window.
2. Look down the pointer at the ring corresponding to the value of the dealer's up card.
3. Observe whether or not the sector bracketed by the pointer window and correct ring is filled, and obey the appropriate legend in the corner of the device.

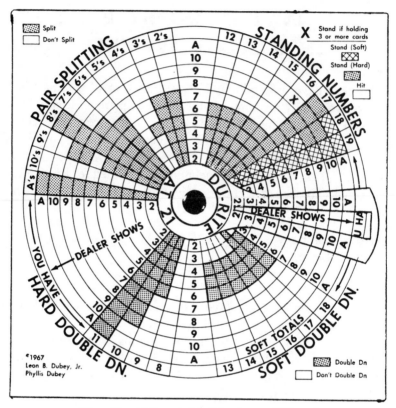

The Du-Rite at 21 Strategy Wheel

There are four basic decisions to be made in casino blackjack. Consequently, the wheel is divided into four sections, each of which corresponds to a particular type of decision.

The numbers in the outermost ring (such as 12, 13, etc. under "standing numbers") correspond to the total for your current hand. Remember, aces count as either 1 or 11, at your option. Face cards count 10, and all others tally at their face, or "pip" value.

The ring numbers 2, 3, 4 . . . 10, A that radiate outward from the center correspond to the dealer's up card. In other words, if his up card is a two, your "decision" will be found in the innermost ring. Conversely, if he shows an ace the correct play will be found in the outside ring labeled A. Notice that these rings appear on the pointer as well.

The crosshatched sectors represent those "soft" totals when you should stand with what you have. You should, of course, stand on equivalent hard totals. Thus the wheel indicates that you should stand on *all* soft totals of 19 or more, and *all* hard totals of 17 or higher. Obviously, all totals less than 11 should be hit since 21 cannot possibly be exceeded.

Priority of Decisions

Your decisions should be made in the following order of priority: (1) Split the pair? (2) Double down? (3) Hit or stand?

Hints

The following hints will enable you to make almost all decisions with one single dialing of the wheel.

Pairs

If the wheel does not call for splitting a pair of sevens or less, hit automatically. Stand on all higher pairs when you do not split.

Doubling Down

Do not bother to check for hard doubling unless your hand total is 9, 10, or 11. Proceed immediately to the standing numbers section of the wheel. If you do have a hard 9, 10, or 11, and the wheel tells you "do not double down," hit automatically.

Soft Double Down

If the wheel does not call for doubling on your soft total of 17 or less, hit automatically.

BIBLIOGRAPHY

Alexander, Howard W. *Elements of Mathematical Statistics*. New York: John Wiley & Sons, 1961.

Brownlee, K. A. *Statistical Theory and Methodology in Science and Engineering*. New York: John Wiley & Sons, 1960.

Cardano, Gerolamo. *The Book on Games of Chance*. Translated by Sydney Henry Gould. New York: Holt, Rinehart & Winston, 1961.

Count of Monte Carlo. *Blackjack—A Little Knowledge Is a Dangerous Thing*. Ojus, Fla.: President Books, 1963.

Crawford, John R. *How to be a Consistent Winner in the Most Popular Card Games*. Garden City, N.Y.: Doubleday & Co., 1961.

Dana, John. *Black Jack—How to Win the Las Vegas Way*. Las Vegas, Nev.: Gambling International, 1965.

Dubins, Lester E., and Savage, Leonard J. *How to Gamble if You Must—Inequalities for Stochastic Processes*. New York: McGraw-Hill Book Co., 1965.

Feller, William. *An Introduction to Probability Theory and Its Applications*. 2nd ed. New York: John Wiley & Sons, 1957.

Goodman, Mike. *How to Win at Cards ("21" & Poker), Dice, Races, Roulette*. Los Angeles: Holloway House Publishing Co., 1963.

Hart, Jack. *Gamble and Win*. Hollywood, Calif.: Onsco Publications, 1963.

Hervey, George, F. *A Handbook of Card Games*. London: Paul Hamlyn, 1963.

Jacoby, Oswald. *How to Figure the Odds*. Garden City, N.Y.: Doubleday & Co., 1947.

Jacoby, Oswald. *Oswald Jacoby on Gambling*. Garden City, N.Y.: Doubleday & Co., 1963.

Lemmel, Maurice. *Gambling: Nevada Style*. Garden City, N.Y.: Doubleday & Co., 1964.

Loeve, Michel. *Modern Probability Theory and Its Applications*. New York: D. Van Nostrand Co., Inc., 1960.

Morehead, Albert H., ed. *The Official Rules of Card Games—Hoyle Up-to-Date*. Racine, Wis.: Whitman Publishing Co., 1959.

Parzen, Emanuel. *Modern Probability Theory and Its Applications*. New York: John Wiley & Sons, 1960.

Radner, Sidney H. *How to Spot Card Sharps and Their Methods*. New York: Key Publishing Co., 1957.

Radner, Sidney H. *The Key to Roulette, Blackjack, One-Armed Bandits*. New York: Ottenheimer Publishers, 1963.

Reid, Ed, and Demaris, Ovid. *The Green Felt Jungle*. New York: Pocket Books, 1964.

Revere, Lawrence. *Playing Blackjack as a Business*. Las Vegas, Nev.: Paul Mann Publishing Co., 1969.

Riddle, Major A., as told to Hyams, Joe. *The Weekend Gambler's Handbook*. New York: The New American Library, 1967.

Scarne, John. *Scarne's Complete Guide to Gambling*. New York: Simon & Schuster, 1961.

Smith, Harold S. *I Want to Quit Winners*. Englewood Cliffs, New Jersey: Prentice-Hall, 1961.

Spitzer, Frank. *Principles of Random Walk*. New York: D. Van Nostrand Co., 1964.

Thorp, Edward O. *Beat the Dealer*. New York: Vintage Books, 1966.

Turner, Wallace. *Gambler's Money*. New York: The New American Library, 1965.

Williams, John D. *The Compleat Strategist*. New York: McGraw-
   Hill Book Co., 1954.
Wilson, Allan. *The Casino Gambler's Guide*. New York: Harper &
   Row, 1965.